FLORA & FAUNA HANDBOOK NO. 1

THE SEDGE MOTHS

OF NORTH AMERICA

(Lepidoptera: Glyphipterigidae)

FLORA & FAUNA HANDBOOKS

This series of handbooks provides for the publication of book length working tools useful to systematists for the identification of specimens, as a source of ecological and life history information, and for information about the classification of plant and animal taxa. Each book is separately numbered, starting with Handbook No. 1, as a continuing series. The books are available by subscription or singly.

Each book treats a single biological subject. Books about the systematics of a single group of organisms (e.g., family, subfamily, single genus, etc.) are scheduled for this series. Books on the ecology of certain organisms or certain regions are planned. Catalogs and checklists of groups not covered in other series will be included here.

The books are complete by themselves, not a continuation or supplement to an existing work, or requiring another work in order to use this one. The books are comprehensive, and therefore, of general interest.

This first book in the series, **The Sedge Moths**, treats the species of these moths that occur in North America and includes information about the classification of the family throughout the world.

Books scheduled in this series:

Handbook No. 1 - THE SEDGE MOTHS, by John B. Heppner
Handbook No. 2 - INSECTS AND PLANTS, by P. Jolivet
Handbook No. 3 - THE POTATO BEETLES, by Richard L. Jacques

THE SEDGE MOTHS OF

NORTH AMERICA

(Lepidoptera: Glyphipterigidae)

JOHN B. HEPPNER

Curator of Lepidoptera and Immatures
Florida State Collection of Arthropods
Center for Arthropod Systematics

CRC Press
Taylor & Francis Group
Boca Raton London New York

CRC Press is an imprint of the
Taylor & Francis Group, an **informa** business

CRC Press
Taylor & Francis Group
6000 Broken Sound Parkway NW, Suite 300
Boca Raton, FL 33487-2742

© 1985 by John B. Heppner
CRC Press is an imprint of Taylor & Francis Group, an Informa business

No claim to original U.S. Government works

Printed on acid-free paper
Version Date: 20150130

International Standard Book Number-13: 978-0-916846-32-9 (Paperback)

Visit the Taylor & Francis Web site at
http://www.taylorandfrancis.com

and the CRC Press Web site at
http://www.crcpress.com

CONTENTS

PREFACE

The revision of the North American sedge moths, family Glyphipterigidae, presented herein is the culmination of a number of years of study on the taxonomy and biology of the included 36 species. Research for the revision was conducted primarily as part of the requirements for the Doctor of Philosophy degree at the University of Florida, Department of Entomology and Nematology, awarded in 1978.

Among the numerous colleagues noted in the section on acknowledgments, two should be singled out here for special mention: Dr. D. H. Habeck, University of Florida, who provided guidance on various aspects of the study, particularly in terms of biologies; and Dr. A. Diakonoff, Rijksmuseum van Natuurlijke Historie, Leiden, Netherlands, who continually provided taxonomic and nomenclatural advice for the family Glyphipterigidae and groups previously included therein.

Although this revision involves Nearctic sedge moths, the text includes host plant data where known for all species of Glyphipterigidae, which, however, does not include very many of the 343 species in the world. Thus, considerable work remains to be done on the biologies of these moths, not to mention the further discovery and description of numerous new species around the world. A large contingent of Glyphipterigidae feed on various sedges and other monocots, thus, providing the derivation of the common name of the family as sedge moths.

Owing to the presence of sedge moths in many riparian ecosystems, these moths are another group of insects that ecologists need to consider in their studies. It is hoped that the present revision of the North American sedge moths will allow relatively easy and certain identification of the North American species for all interested persons.

Gainesville   J. B. Heppner
December 1984

vii

INTRODUCTION

The present study of the Glyphipterigidae[1] was under-
taken to clarify the status of the taxa included in the older
concept of Glyphipterigidae (**auctorum**) and as a revision of
the North American fauna. It represents the first comprehensive study of these moths in North America north of
Mexico using modern systematic procedures involving assessment of a number of morphological and biological characters in relation to their phylogenetic status. Although not
fully documented herein, this study also represents a world-
wide generic study made in conjunction with the revision of
the North American species, primarily to ascertain realistic
generic limits and the placement of various genera once
included in the heterogeneous concept of the family. Extralimital genera are noted in a catalog of generic names
associated with the Glyphipterigidae (Heppner, 1982b).

This study originated in 1972 upon the suggestion of W.
Donald Duckworth, Smithsonian Institution, as a family need-
ing revision not only for the North American species but also
in regard to its position within the Ditrysia. A previous re-
port involved the higher category classification of the group,
the segregation into unrelated families (Choreutidae, Glyphipterigidae, and Immidae), and relationships within the
microlepidopterous Ditrysia (Heppner, 1977). The "Atychiidae," a west Palearctic family until now and formerly included in Glyphipterigidae, is included in Sesioidea (as the
family Brachodidae), in addition to Sesiidae and Choreutidae,
in a review of sesioid classification (Heppner and Duckworth,
1981). Many of the genera thought to be in "Choreutinae,"
yet possessing a naked haustellum (true Choreutidae have a
scaled haustellum), have been transferred to an enlarged
concept of "Atychiidae," with a name change to Brachodidae
due to nomenclatural problems involving the type-genus **Atychia** Latreille (Heppner, 1979).

In the past, the Glyphipterigidae (**sensu lato**) encom-
passed approximately 1000 species throughout the world.

---

[1] The spelling of Glyphipterigidae is based on Art. 29(d)
of the International Code of Zoological Nomenclature, as derived from **Glyphipterix** Hübner, rather than as Glyphipterygidae, based on the invalid emendation **Glyphipteryx** (**auctorum**)

The family thus defined, mainly due to the work of E. Meyrick, formed one of the most heterogeneous conglomerations of unrelated genera and species ever assembled in the last century. Although Meyrick did extensive work on the microlepidoptera, describing over 14,000 species from 1875 until 1939, he unfortunately included species in Glyphipterigidae mainly on superficial similarities. The taxa he included in his concept of Glyphipterigidae are presently assigned to no less than 25 distinct families among several superfamilies. Over 230 of the included species are in the genus **Imma,** now assigned to the new family Immidae (Heppner, 1977). The Choreutidae are restricted to taxa formerly grouped as the subfamily Choreutinae. The present world fauna of Choreutidae encompasses 18 described genera and 377 species (includes species to be added to the Palearctic by Diakonoff (in press) and to the Nearctic fauna in a forthcoming revision) (Heppner, in prep. b). The Glyphipterigidae (**sensu stricto**) are currently restricted to a world fauna of 26 genera and 343 species, including the new taxa described herein. The enlarged concept of the Brachodidae encompasses 11 genera and 96 species (Heppner and Duckworth, 1981).

Some species of Choreutidae and Glyphipterigidae are superficially similar in metallic iridescent wing maculation, possibly due to mutual adaptions to diurnal adult activity, but they do not have a recent common ancestor, as demonstrated by evolutionarily conservative and more fundamental characters, notably the abdominal articulation, larval chaetotaxy, and pupal morphology (Brock, [1968]; Heppner, 1977). The Glyphipterigidae are now placed in the superfamily Copromorphoidea and the Choreutidae are placed in the Sesioidea. The placement of other genera associated with the old concept of Glyphipterigidae is noted in the generic catalog (Heppner, 1982b), while certain North American taxa have already been excluded (Diakonoff, 1977a; Heppner, 1978) and are noted on page 21.

Currently there are 5 genera and 36 species of Glyphipterigidae from North America north of Mexico, of which 1 genus and 22 species are newly described herein. In North America the percentage increase of new species is much larger in Glyphipterigidae than recent revisions have produced for other families of microlepidoptera. The percentage of previously undescribed species comes to 70% in Glyphipterigidae, while in the recent Nearctic revisions for Ethmiidae (now a subfamily of Oecophoridae) (Powell, 1973) it is 18%; for the first five subfamilies of Pyralidae (Munroe, 1972-[1974], 1976) it is a consistent 16%; and for Oecophoridae exclusive of Ethmiinae (Hodges, 1974) it is only 10%. However, in a similar first revision being completed for the Nearctic Choreutidae (Heppner, in prep. b) the percentage is 41% and in another first revision being prepared on Nearctic Argyresthiidae by Paul A. Opler, it probably will be closer to 100%. In a first revision among microlepidoptera it is not too surprising that the percentages are so high. In Glyphipterigidae and Choreutidae, especially so, since the moths are diurnally active and small diurnal microlepidoptera are generally less collected than nocturnal moths. The Palearctic fauna comprises a total of about 100 species for both families, but at this writing it is not yet known what the final percentage of new species will be in the concurrent Palearctic revision by Diakonoff (in press). There is only one Holarctic species known and this is in the only Holarctic genus, Glyphipterix.

Glyphipterigidae are relatively small moths, ranging from 4.5-30 mm in wingspread. The largest known species are in the genus Sericostola, only recently added to the family (Heppner, 1984). Larvae of Glyphipterigidae are borers of seeds, terminal buds, stems or culms, and leaf axils, predominately in monocotyledonous plant families, especially Cyperaceae and Juncaceae. They rarely are leaf miners (1 species). The species generally prefer damp or marshy habitats, areas where the two most widely used plant families are common. Wherever the hosts are successful various species of glyphipterigids can be found, often in riparian habitats in areas that appear to be too dry or at too high an altitude to support the species beyond the humid microhabitat. Tropical species appear to also favor monocotyledons as hosts, although some species utilize hosts in other plant groups.

The biological information presented in the text is based on both literature reports and field work, either my own or that of colleagues. Several grants allowed field studies in most of the eastern United States and several western states from 1973 to 1977. Since these studies were usually of relatively short duration, considerable biological research is still required. The majority of the species remain unknown biologically.

A number of institutional and private collections provided the adult specimens upon which this revision is based. These collections include both North American and European collections (see abbreviations under Specimens Studied and Taxonomic Procedures). My own field studies augmented the number of available specimens of several species. Although many species are adequately represented in accumulated collections, a number of species have small series or only a single example. Where characters denoted valid species, these were described even if represented by only a single specimen.

Several areas in North America are very poorly represented by available specimens and are the most promising areas for future field studies. The southeastern states and the western regions of both the United States and Canada contain diverse habitats potentially harboring a greater number of species than described herein for Glyphipterigidae. Surprisingly few records are available for New England or the mid-Atlantic states and those that are available often are from 75-100 years old. Arizona has a number of species that are based on only one or a few specimens; a similar situation exists in the northern Rocky Mountains. Thus, despite the high percentage of new species in this study, this revision is only a beginning until North America is much better surveyed for microlepidoptera.

In the bibliographic synonymies, all known references are included where specific names have been cited. For genera, all known references are cited except for the large genus **Glyphipterix** where only those references are included that involve a discussion of the genus or where two or more species are noted. For the family references, most known references are included that discuss the family, use the family name in conjunction with two or more species, or use the name in a checklist or catalog. The revision involved the North American fauna; to have made an exhaustive survey of all the world literature for references of scientific names in the numerous, often obscure, checklists of local faunas over the past 150 years would have been beyond any usefulness for this study. Consequently, the bibliographic references for the family usage, for **Glyphipterix**, and for the single Holarctic species, **Glyphipterix haworthana** (Stephens), are incomplete. Even so, all major references are included and the percentage of inclusiveness is quite high. The reader is referred to the forthcoming revision of the Palearctic Glyphipterigidae (**sensu lato**) by Diakonoff (in press) for references to many of these local faunal lists.

The following new taxa and nomenclatural changes are included in the present revision: 1 new genus, 7 new species groups, 22 new species, 3 new subspecies, and 6 new lectotypes.

ACKNOWLEDGMENTS

Research for this revision was initially conducted at the University of Florida, Gainesville. The Department of Entomology and Nematology, Institute of Food and Agricultural Sciences, University of Florida, provided support and facilities as well as partial support for field research in Florida and the southeastern United States. Additional biological information and specimens were obtained through other projects conducted under a grant from the Florida Department of Natural Resources (Dale H. Habeck, principal investigator).

The National Science Foundation provided a dissertation enhancement grant (DEB76-12550) in 1976 for field research in the western United States and travel funds for research at several eastern North American museums and museums in Europe, primarily the British Museum (Natural History), London. W. Donald Duckworth kindly provided partial additional funding from the Smithsonian Institution for visits to other European museums in early 1977 in conjunction with the trip to London.

Biological studies in 1975 in the Everglades National Park, Florida, were facilitated by C. McClain, Park Superintendent, who issued a collecting permit. The late Richard Archbold kindly provided a grant for ten days research at the Archbold Biological Station, Lake Placid, Florida, in 1975.

A pre-doctoral fellowship awarded by the Smithsonian Institution in 1976 allowed extended study of types and other specimens at the National Museum of Natural History, Washington, D.C. W. Donald Duckworth made arrangements to provide needed space and equipment. The Smithsonian Institution fellowship award also funded part of the costs of travel to the British Museum (Natural History), London, to study types.

The following individuals and their respective institutions kindly provided the specimens upon which this revision is based: Paul H. Arnaud, Jr. (California Academy of Sciences, San Francisco), André Blanchard (Houston, Texas), Vernon A. Brou (Edgard, Louisiana), Wolfgang Dierl (Zoologische Sammlungen des Bayerischen Staates, Munich, West Germany), R. B. Dominick (McClellanville, South Carolina), Julian P. Donahue (Los Angeles County Museum of Natural History, Los Angeles, California), W. D. Duckworth (Smithsonian Institution, Washington, D.C.), Thomas D. Eichlin

(California Department of Food and Agriculture, Sacramento), Kenneth E. Frick (Mississippi State University, Starkville), K. C. Kim (Frost Entomological Museum, Pennsylvania State University, University Park), Charles P. Kimball (West Barnstable, Massachusetts), Robert E. Lewis (Iowa State University, Ames), Bryant Mather (Clinton, Mississippi), Eugene G. Munroe (Canadian National Collection, Agriculture Canada, Ottawa), Daniel Otte (Academy of Natural Sciences, Philadelphia, Pennsylvania), L. L. Pechuman (Cornell University, Ithaca, New York), Jerry A. Powell (Essig Museum of Entomology, University of California, Berkeley), Frederick H. Rindge (American Museum of Natural History, New York), Gaden S. Robinson and Klaus Sattler (British Museum (Natural History), London, United Kingdom), Vincent D. Roth (Southwestern Research Station, Portal, Arizona), R. E. Silberglied (Museum of Comparative Zoology, Harvard University, Cambridge, Massachusetts), William J. Turner (Washington State University, Pullman), Howard V. Weems, Jr. (Florida State Collection of Arthropods, Florida Department of Agriculture and Consumer Services, Gainesville), R. L. Wenzel (Field Museum of Natural History, Chicago, Illinois), and Barry Wright (Nova Scotia Museum, Halifax, Nova Scotia, Canada).

Several other institutions provided specimens that belong exclusively to the now excluded Choreutidae. These institutions will be acknowledged in the revision of Nearctic Choreutidae now in final preparation.

Special thanks are due for several persons who made efforts to collect moths for me: G. B. Fairchild, Harold N. Greenbaum, and Kenneth J. Knopf (all University of Florida, Gainesville), and W. H. Pierce (Florida Department of Agriculture and Consumer Services, Gainesville). Jerry A. Powell made several important additions to the material from the University of California collection by collecting additional specimens from several western states during annual field trips.

Plant identifications were kindly made by David Hall (University of Florida Herbarium, Gainesville) and Stanwyn G. Shetler (Smithsonian Institution, Washington, D.C.). Victor E. Krantz, assisted by Harold Dougherty, of the Smithsonian Institution Photographic Unit (USNM), kindly photographed genitalia and adults for this revision. The scanning electron micrographs were made by the Smithsonian Institution Natural History Museum SEM Laboratory.

Several persons provided helpful criticism of the manuscript or discussion of taxonomic problems: J. F. G. Clarke,

Donald R. Davis, W. Donald Duckworth, W. D. Field, Ronald W. Hodges, and E. L. Todd (all of the Lepidoptera Section, National Museum of Natural History, Washington, D.C.), and Dale H. Habeck (University of Florida, Gainesville), K. Sattler (British Museum (Natural History), London), A. Diakonoff (Rijksmuseum van Natuurlijke Historie, Leiden, Netherlands), and Niels P. Kristensen (University of Copenhagen, Denmark).

The critical reading of this manuscript and the many helpful comments by Thomas E. Emmel, Jonathan Reiskind, and Minter J. Westfall, Jr., Department of Zoology, and Dale H. Habeck, Carol A. Musgrave, and Robert E. Waites, Department of Entomology and Nematology, University of Florida, Gainesville, are gratefully acknowledged.

A. Diakonoff and I had many hours of discussion and extensive correspondence during the course of this revision owing to his concurrent study of the Palearctic fauna, and for all his helpful comments I am most grateful.

TAXONOMY

Historical Review

The original conception of glyphipterigid and choreutid moths was closer to our present classification in many ways than that used for most of the current century. In the early 1800s these moths were grouped in Tineina, with the glyphipterigid genera near the tineoid moths and the choreutid genera near the tortricids. Such genera as "**Atychia**," included in the heterogeneous concept of Glyphipterigidae by Meyrick (1914c), were originally grouped among what were called crepuscular moths and, in fact, relatively closely to the Sesiidae, which are now considered to be only very advanced and specialized relatives of **Brachodes (=Atychia)** ancestors. Other genera, like **Imma**, were also classified elsewhere, but until the 1880s a number of classifications placed the glyphipterigid genera among the tineoid moths and the choreutid genera among the tortricoid moths (Zeller, 1847; Stainton, 1854b; Heinemann, 1870; Snellen, 1882). Often these genera were not recognized in distinct families but only grouped within a large assemblage called Tineina, what would now be called microlepidoptera exclusive of Cossoidea, Pyraloidea and Zygaenoidea.

A family concept for the Glyphipterigidae (sensu stricto) was first used by Stainton (1854b) but based on the emended spelling of **Glyphipterix** Hübner as **Glyphipteryx (auctorum)** (Curtis, 1827 and Zeller, 1839a, made the first emendations of the generic name). Although Ford (1949) was the first recent author to revert to the family spelling based on the original spelling of the type-genus, it was Rosenstock (1885) who actually first used the spelling Glyphipterigidae in an obscure paper on Australian moths. Meyrick used the emended spelling, and, consequently, so did most other authors until recently. Our zoological code now requires genera to be spelled as originally spelled, thus, Glyphipterigidae is currently used. The family name has only one synonym, Aechmidae.

Prior to the discovery of new tropical and southern hemisphere species, most glyphipterigids were described in **Glyphipterix** (virtually always in the emended form **Glyphipteryx**). Up until 1870 there were only 13 described species in **Glyphipterix** for the world and this was the only genus (Stainton, 1870). Before the 1850s the genera **Aechmia** and **Heribeia** were used in addition to **Glyphipterix**, but these names were later synonymized with **Glyphipterix**. Meyrick combined the glyphipterigids and choreutids as closely related moths by the late 1880s but retained them in his broad concept of Plutellidae, what is now approximately equivalent to our Yponomeutoidea. He still retained them in Plutellidae in his first **Handbook of British Lepidoptera** (Meyrick, 1895), one of the few works where he elaborated on his ideas of higher category classification. He listed the moths as the family Glyphipterigidae by the time he completed his part of the **Lepidopterorum Catalogus** (Meyrick, 1913b), likewise followed in his Glyphipterigidae part of the **Genera Insectorum** (Meyrick, 1914c). He originally placed Imma in Yponomeutidae in his revision of the genus (Meyrick, 1906) but, along with many other unrelated genera, placed them in his Glyphipterigidae in the catalog (1913b) and generic revision (1914c). Thereafter, Meyrick and others described several hundred new species and many new genera as Glyphipterigidae (sensu Meyrick). Most authors still grouped all the major genera Meyrick placed in his concept of the family in Glyphipterigidae until the 1960s.

Only relatively recently has Meyrick's concept of the family been questioned. The initial research was conducted by J. P. Brock and published as a short note on the dissimilar morphology of the Glyphipterigidae and the Choreutidae (Brock, [1968]). His survey of the morphological features of the Ditrysia (Brock, 1971)—a very significant paper which did not until recently become fully integrated into our classification of the Lepidoptera—also noted the adult morphological features by which the two families are unrelated phylogenetically.

The research presented in the present revision, in particular the study of the genera formerly assigned to the Glyphipterigidae, confirms much of Brock's work and clearly demonstrates the distinct phylogeny of the Glyphipterigidae and Choreutidae. The basic features of larval chaetotaxy, pupal morphology and behavior, and basic adult morphology, particularly the abdomino-thoracic articulation, demonstrate that the two families are unrelated and even belong to different superfamilies as this higher category is currently defined. The distinctions between the two families and their presumed relationships to other families were elaborated previously (Heppner, 1977).

The present restriction of Glyphipterigidae to 26 genera world-wide still leaves about 200 generic names (including misspellings) that have at some time been associated with the heterogeneous concept of the family. North American genera and species excluded from the treatment herein are noted in the section on excluded taxa (page 21). The catalog of excluded generic names treats the other names also (Heppner, 1982b).

Specimens Studied and Taxonomic Procedures

This revision involved the study of about 1500 adult specimens for the 36 species listed. Most of the specimens were accumulated through the kind cooperation of a number of North American and European institutions and private collections. The following abbreviations are used to identify these various collections:

AB	André Blanchard Collection, Houston, Texas
AMNH	American Museum of Natural History, New York, New York
ANSP	Academy of Natural Sciences, Philadelphia, Pennsylvania
BM	Bryant Mather Collection, Clinton, Mississippi
BMNH	British Museum, (Natural History), London, United Kingdom

CAS	California Academy of Sciences, San Francisco, California
CDAS	California Department of Food and Agriculture, Sacramento, California
CNC	Canadian National Collection, Agriculture Canada, Ottawa, Canada
CPK	Charles P. Kimball Collection, West Barnstable, Massachusetts (now at MCZ)
CU	Cornell University, Ithaca, New York
DFB	Dale F. Bray, University of Delaware, Newark, Delaware
FEM	Frost Entomological Museum, Pennsylvania State University, University Park, Pennsylvania
FMNH	Field Museum of Natural History, Chicago, Illinois
FSCA	Florida State Collection of Arthropods, Gainesville, Florida
ISU	Iowa State University, Ames, Iowa
JBH	John B. Heppner Collection, Gainesville, Florida
LACM	Los Angeles County Museum of Natural History, Los Angeles, California
MCZ	Museum of Comparative Zoology, Harvard University, Cambridge, Massachusetts
MSU	Mississippi State University, Starkville, Mississippi
NSM	Nova Scotia Museum, Halifax, Nova Scotia, Canada
RBD	R. B. Dominick Collection, McClellanville, South Carolina
SWRS	Southwestern Research Station, Portal, Arizona
UCB	University of California, Berkeley, California
USNM	National Museum of Natural History, Smithsonian Institution, Washington, D.C,
VAB	Vernon A. Brou Collection, Edgard, Louisiana
WSU	Washington State University, Pullman, Washington
ZSBM	Zoologische Sammlungen des Bayerischen Staates, Munich, West Germany

A trip to the British Museum (Natural History) and other European museums, allowed an extensive search for all North

American specimens in most of the major European museums. In addition to the North American material, several
hundred exotic species were also examined, in varying de-
grees of completeness, to ascertain generic limits and the
proper placement of many of these species. The generic
studies allowed relatively definitive species totals to be
noted for each genus in lieu of complete revisions for each
genus.

In addition to specimens accumulated from institutions,
several hundred specimens were collected on field trips from
1973-1977, involving many new range extensions and additions of longer series for some species as well as some new
species. These field trips also allowed biological observations for several species not occurring in Florida, but the
studies remain incomplete except for one species relatively
well-studied in Florida. The biologies of 6 glyphipterigids
have been elucidated to some extent but only two species
have been successfully reared. One species of **Glyphipterix**
from Japan (**G. semiflavana** Issiki) has a larval and pupal
description with accompanying illustrations (after Moriuti,
1960) included because of the lack of available immature
stages for even a single North American species of this
genus. Most of the North American species remain to be
investigated biologically.

Morphological Studies

Preparation of the genitalia slides follows the techniques
described by Clarke (1941) and Robinson (1976). Whole
mounts of adult heads, wings, and larvae were prepared using
the same procedures as for the genitalia slides except that
the dissections varied for each type of anatomical part: wing
slides were usually mounted dry under a sealed cover glass.
Chlorazol black was the usual stain, inasmuch as there is a
useful differentiation between sclerotized areas (which are
not stained and remain brown) and unsclerotized areas (which
are stained blue). Canada balsam was used as the mounting
medium for most slides, except those made in London at the
British Museum (Natural History), where euparol is preferred.
Whole mounts of larvae were made by ventrally slitting the
pelt prior to mounting in balsam. The pelt was mounted flat
on a slide with the head removed and mounted upright; the
left mandible and submentum were removed for separate
mounting on the same slide. The larval head was drawn
laterally prior to upright mounting in balsam. Usually a
depression slide was used for the larval slides so the head

could be easily accommodated in the depression, with the pelt mounted to the right of the depression under a second, often rectangular, cover glass.

The heads for the morphological drawings were handled as whole mounts using procedures similar to those outlined above. Heads for the scanning electron micrographs were each affixed to a small plastic cube with white wood glue, which was then glued onto a cover glass, whereupon the whole assemblage was given a relatively heavy coating (actually 2-3 separate coatings) of gold prior to use in the scanning electron microscope at the National Museum of Natural History, Smithsonian Institution. Some difficulty was experienced with the fully scaled heads since the gold film does not penetrate the interstices of overlapping scales. Without a nearly 100% gold coating heads tend to develop "charging" which results in poor photographs.

Measurements of adult features were made with a micrometer disc in a stereomicroscope. When available, several individuals of each sex were used for measurements. The number of genitalic preparations usually involved several specimens of each sex from several localities over the range of the species as specimens were available; the number of slides studied is noted after the description of the genitalia for each species. Head and wing mounts were made only for the generic type-species, although wing venation was studied in situ for each species as well. Drawings and genitalia photos were made from particular slides; the slide number and ownership are noted in the figure captions. The figures were generally reproduced at a uniform size in each sequence to facilitate grouping of the figures and to allow uniform visualization of details.

Descriptions

Inasmuch as these are generally rather obscure moths and inadequately described in earlier literature for those species previously described, all the included species are given a full description. The descriptions are generally composite in being inclusive of variations found among available specimens of each species, but were initially composed by the use of the best available male and female specimens. Colors are as seen under incandescent illumination under a steriomicroscope. Flown specimens tend to be brown in areas that are dark fuscous in fresh specimens, undoubtedly due to fading as a result of exposure to direct sunlight. This factor of color fading must be taken into account when identifying specimens.

The adult specimens photographed for the illustrations were labelled with a Smithsonian Institution label reading, "Photograph on file USNM" in blue ink; the negatives are in the National Museum of Natural History, Smithsonian Institution, Washington, D. C. The genitalia photographs are identified by slide numbers. Wing maculation of both sexes of the adult are illustrated when available, even though most are not markedly dimorphic.

Locality data is tabulated after the descriptions for each species. Lectotypes are designated for a few of the species. Where information is not noted in the locality data (e. g., no date, no year, or no collector name, etc.) the information is unknown. Data in brackets ([ ]) are insertions of known information derived from other sources, either original descriptions, maps, or subsequent papers (e. g., Essig (1941) on Walsingham's California and Oregon itineraries for 1871–1872). However, county names are added to all localities when possible even though many of these are not actually noted on specimen labels; these county names are not in brackets if absent from the labels so the data sequences are not made inconsistent with other data in the same county where the county name is on the specimen label.

At the end of the sections on type specimens (holotypes and paratypes) for each species, collections are noted which will receive some or all of these specimens: thus, when possible, paratypes have been selected for deposition at least in the British Museum (Natural History), Canadian National Collection, Florida State Collection of Arthropods and the Smithsonian Institution. All holotypes from the University of California, Berkeley, are deposited on indefinite loan with the California Academy of Sciences, as is their custom.

Taxonomic Procedures

A combination of modern systematic approaches has been used in assessing the limits of genera and species and in determining their relationships. A strict cladistic approach has not been used inasmuch as I consider the methodology espoused by Hennig (1950) to encompass several errors, notable among which is the lack of consideration for the often exceptional results of differential rates of evolution. Thus, genera are not dichotomously split on the basis of recency of an inferred common ancestor as the only criterion for such splitting, as proposed by Hennig. In general, this revision is

based on the principles of systematics as elaborated by Mayr
(1969). Thus, many characters of adult head morphology,
wing venation, genitalia features, and morphology of immature stages (when known), are used in assessing generic and
species limits herein. Such a taxonomic procedure has been
found useful and authoritative in other families of Lepidoptera as well as throughout the animal kingdom.

The Hennigian concepts of relationships demonstrated by
shared apomorphic characters and shared relationships in
plesiomorphic character states were also used in this revision. Thus, taxa are arranged in the text from less
advanced forms to more highly derived forms to the extent
possible. Genera having more generalized morphological
features inferred to be less advanced, are placed at the
beginning of the treatment, while genera possessing characters inferred to be more specialized are placed toward the
end. Such a treatment is similarly used for the species of
each genus and follows the pattern now adopted in all families of Lepidoptera.

Initial groupings of apparent species early during the
revision resulted in considerable lumping of what later were
discerned to be several species, mainly in **Glyphipterix**.
There are a number of "sibling species" complexes in **Glyphipterix** world-wide, several in the Nearctic. In almost all
cases the genitalia show discernible and consistent characters
distinguishing the different species, while the wing maculation is often virtually identical or initially confusing until
specimens are compared. Superficially some species of **Glyphipterix** are also similar to species herein segregated into
another genus, **Diploschizia**. In most species either sex
offers distinguishing characters for the species, although the
male genitalia often offer characters that are easier to use
in identification of unknown specimens.

The species complexes identified in this revision cannot
by definition be termed true sibling species, since discernible
morphological distinctions exist. Sibling species by definition
are exactly identical phenotypically yet reproductively
isolated even if sympatric. One could possibly call these
complexes superspecies and the members semispecies but the
actual definitions of these terms allude to as yet incomplete
separation of the semispecies as true species, that is, fully
reproductively isolated. It is not known whether any hybridi-
zation occurs between any of the species of **Glyphipterix** but
from the characters used to define the species herein, they
are considered full species. I prefer, therefore, to use the

term "species complex" in discussing the similar appearing species found in **Glyphipterix** and **Diploschizia**.

Many of the species in these complexes have genitalia differences that clearly distinguish them as reproductively isolated. Thus, it is especially from such species that exhibit large and unquestionable specific differences yet often having wing maculation similar to species exhibiting little genitalic differences, that are used to correlate the significance of the apparently minor differences in the more closely related species. Where the genitalia are identical in every feature and the populations of a species are especially allopatric or represent a distinct phenotype, I have sparingly used the subspecific category so the disjunct populations can be easily identified in discussions.

The adults of Glyphipterigidae are quite localized and usually remain in close proximity to their host plants. Consequently, some members of the species complexes may be only recently isolated, although many of the members are sympatric. The species complexes are mainly in the montane regions of western North America where habitat isolation is often so distinct and where extensive speciation has occurred. Host plant specificity undoubtedly also had a major role to play in this speciation.

The use of more formalized species-groups in **Glyphipterix** is an innovative feature of this revision. The species of the genus have been grouped into ill-defined assemblages in the past, primarily in the Palearctic by Meyrick (1914c). Species-groups have been successfully used in such diverse world-wide genera as **Ethmia** (Oecophoridae: Ethmiinae), **Depressaria** (Oecophoridae: Depressariinae), **Stigmella** (formerly **Nepticula**) (Nepticulidae), **Coleophora** (Coleophoridae), **Pyrausta** (Pyralidae), and others. Since only the Nearctic **Glyphipterix** have thus far been newly segregated into species-groups, some of the groupings may require alteration as exotic species are integrated into this classification of the genus. The species-groups used in this revision are used for convenience and for grouping relatively closely related species, and are not to be considered equivalent to subgenera, although the latter category may eventually be considered for a division of the genus. In addition to allowing clustering of related species, a useful secondary feature of using species-groups is that relationships can be indicated for extra-limital species, as is the case with one North American species placed in the **circumscriptella** species-group which has numerous Neotropical adherents. The species-groups are not exhaustively defined but are given a detailed diagnosis for characters held in common among the included species.

Systematics and Phylogeny

Systematic Position of the Glyphipterigidae

Subsequent to the early classifications which placed the group in the Tineina moths, as previously noted, the glyphipterigids were generally associated with yponomeutoid moths by most researchers, primarily due to superficial similarities of external head characters, wing maculation and venation, and some similarities in the genitalia. Wing venation was perhaps the main criterion used in associating these moths with the yponomeutoids prior to detailed studies of the larvae and genitalia. Wing venation was the predominant character system used by Meyrick in assessing relationships among the microlepidoptera. Thus, when Meyrick combined the choreutids and glyphipterigids into one heterogeneous family, he aligned them with the Plutellidae, Yponomeutidae, and Heliodinidae.

In Meyrick's last major work where he discussed higher category classification and relationships (Meyrick, 1928), he proposed the superfamily Glyphipterygoidea for three families he excluded from Yponomeutoidea: Glyphipterigidae (including Choreutidae), Heliodinidae, and Heliozelidae. The latter family is completely unrelated and is not even ditrysian, belonging instead to the Incurvarioidea in Monotrysia, but as with the other two was included only on superficial resemblance and what Meyrick thought were some relationships of wing venation. The wing venation is actually very different in these families. Meyrick's Heliodinidae also included a number of genera actually belonging to Oecophoridae (Gelechioidea). In the same work he also defined the Copromorphoidea to include the Copromorphidae, Carposinidae, and Alucitidae. Meyrick's higher classification in this work was largely ignored by his colleagues at that time, although his superfamily innovations have since been found to be valid to some extent.

Since 1928 the Glyphipterigidae (always including the Choreutidae) have become exclusively regarded as yponomeutoid. Following Brock's presentation of the unrelated morphology of Glyphipterigidae and Choreutidae (Brock, [1968]) and his use of the superfamily Aegerioidea (=Sesioidea) for Sesiidae and Choreutidae (Brock, 1971), the glyphipterigids remained in Yponomeutoidea until my recent paper detailing the affinities of the Glyphipterigidae to Copromorphoidea (Heppner, 1977), mainly because Brock's work was not immediately accepted. In my paper (Heppner, 1977) it was noted that major morphological characters of the adult, larva, and pupa demonstrate affinities to Copromorphoidea and not Yponomeutoidea. For the first time the Epermeniidae were also considered to be closely related to Glyphipterigidae, as also to Carposinidae. It is interesting to note that Turner (1947) considered the Glyphipterigidae (sensu lato) to be related both to Sesiidae and Copromorphidae, which indeed is correct in regard to the two separate affinities of Choreutidae and Glyphipterigidae (sensu stricto), respectively.

Wing venation patterns of several families in several superfamilies show apparent similarities, notably in the Glyphipterigidae, Choreutidae and true yponomeutoid moths. As we know, these similarities of wing venation are only similarities and do not necessarily represent especially close relationships.

Taking into account the three major character systems discussed in my earlier paper (Heppner, 1977), a different picture emerges than what was believed by Meyrick and others. The abdominal articulation with the thorax offers an especially useful character in assessing higher category relationships in that it appears not to be significantly affected by the selective pressures among related species and, thus, has been found to be evolutionarily very conservative (Brock, 1971; Heppner, 1977). The ventral portion of this articulation is also easily studied since it can be viewed in every genitalia slide preparation where the abdominal pelt has been mounted along with the genitalia, as is the usual practice. The abdominal articulation is formed by muscles attached to two apodemes at the junction with the thorax, which have developed into two major morphological types: 1) a tineoid rod type (Fig. 128) where the apodeme has an apparent extension as a narrow, long rod extending posteriorly into the 2nd abdominal sternite (usually the apodeme is without lateral sclerotizations); and 2) the tortricoid type (Fig. 129) where the apodeme is usually stouter and usually has distinct lateral sclerotizations at the base, but lacks a distinct extension into the abdominal sternite. All Lepidoptera from Immoidea to the advanced macro-moths and butterflies have the tortricoid type of articulation apodemes, while Yponomeutoidea and all less advanced Lepidoptera have the tineoid type (Brock, 1971). This character clearly distinguishes the Choreutidae, which have a tortricoid type of articulation, from the Glyphipterigidae, which have a tineoid rod type of articulation. Immidae, Brachodidae, and Sesiidae all have the tortricoid type of articulation.

The second major character system noted in the previous paper (Heppner, 1977) involves larval chaetotaxy. Several groups of setae apparently have shifted position or setal number relatively slowly in the evolution of Lepidoptera in general. Shifts that have occurred usually correlate with other characters defining distinct families or superfamilies. Occasional odd families and genera occur but the correlations have been found valid in almost all cases, with the exceptions being very rare, so that larval chaetotaxic shifts are very useful in assessing relationships among higher categories. The prothoracic lateral (L-group) setae are especially useful and often are used for family recognition in keys to Lepidoptera larvae. As noted in the previous paper (Heppner, 1977) Glyphipterigidae, Epermeniidae, and other Copromorphoidea, have a bisetose larval L-group, while in Yponomeutoidea and Sesioidea it is trisetose. A table presented the setal number for other microlepidoptera in the previous paper and illustrated that this character correlates well with other characters distinguishing families and superfamilies. Larval characters also show that the five families now proposed for Copromorphoidea are related, since they all have some development of more dorso-caudally positioned spiracles on the 8th abdominal segment. Although this character is generally prevalent in boring larvae, it is unusually prominent in copromorphoids. Additionally, both Glyphipterigidae and Epermeniidae larvae have unusual protrusions surrounding the spiracles, the largest protrusions being found on the prothorax and 8th abdominal segment.

Lastly, the pupal abdominal spination on the dorsum of the abdominal segments is useful in assessing relationships, since this morphological character correlates with pupal behavior at adult eclosion. The spination is either prominent or absent in different families, virtually always of one kind or the other within one superfamily. Thus, a significant difference is found between Glyphipterigidae and Yponomeutoidea to Sesioidea. The latter two superfamilies have spined pupae (Sesioidea) or a protruded pupa (Yponomeutoidea). The pupa is protruded at adult eclosion in both these superfamilies, while in Glyphipterigidae and other Copromorphoidea, the pupa is unspined and is not protruded at adult eclosion. Plutellidae are unusual in Yponomeutoidea in not protruding the pupa but are related to other yponomeutoids by many other characters. The Plutellidae cannot be included in Copromorphoidea, since they have a trisetose L-group on the larval prothorax and have other characters common to yponomeutoid families. Some recent authors maintain the plutellids as a subfamily of Yponomeutidae (Bradley, 1972; Moriuti, 1977). Further research is needed on the relationships of Plutellidae and Glyphipterigidae since the pupae of the two families are similar, while larval characters, in terms of chaetotaxy and other characters, align the two families to the two superfamilies, Copromorphoidea and Yponomeutoidea. The pupae of Epermeniidae are similar to Carposinidae pupae, whereas Plutellidae and Glyphipterigidae pupae have produced spiracles. Thus, further studies may indicate that either Glyphipterigidae are best considered yponomeutoid (then the only family of Yponomeutoidea with bisetose larvae) or the superfamilies Copromorphoidea and Yponomeutoidea should be merged as one superfamily.

Immidae were aligned with Sesioidea when first named (Heppner, 1977) but I have since concluded that there are two many major character differences between them and other Sesioidea to retain the family in this superfamily. Common (1979) demonstrated that the larvae have some secondary setae, the pupae are without spines and are not protruded at adult eclosion, features uncharacteristic of Sesioidea. In a separate paper on Immidae (Heppner, 1982d) the affinities to Yponomeutoidea are discussed. Immidae now are in their own monobasic superfamily, Immoidea (Common, 1979).

It is conceivable that Meyrick's "Glyphipterygoidea" may eventually be reinstated for Glyphipterigidae and Epermeniidae, but I find no significant characters to exclude these two families from Copromorphoidea and, thus, do not consider their separation as another superfamily to be valid in relation to the limits of other superfamilies in Lepidoptera. The relationships to Plutellidae noted above are further evidence against a third superfamily for this area of the Ditrysia. The two families generally are smaller in size and have different wing shapes than other copromorphoids. However, the generalized species as are found in Copromorphidae are undoubtedly the less advanced elements of the superfamily, indicating a trend from larger species with primitive wing venation (Copromorphidae), to reduction in

veins (Carposinidae), to wing separation (Alucitidae), and to
smaller diurnal moths (Epermeniidae and Glyphipterigidae).

Recent studies by Kuznetsov and Stekolnikov (1976, 1977)
have attempted to deduce phylogenetic relationships among
microlepidoptera solely on male genital musculature. The
results of their studies have indicated relationships between
Glyphipterigidae and Plutellidae and also between Choreutidae and Sesiidae. Although the genital musculature offers
additional character systems for analysis, it seems doubtful
that one should base higher category relationships solely on
one character system when so many others are also available. Reliance on single character systems in the past has
led to many errors. A more recent paper, by Kyrki (1984),
using so-called pleural lobes of the male genitalia, also uses
only a single character system (which is highly subject to
selective pressure at the species level) to transfer a variety
of unrelated families into a new classification of Yponomeutoidea. A holistic approach, using a broad spectrum of
characters is undoubtedly better.

Phylogeny of Glyphipterigidae

The basic characters held in common among the five
families of Copromorphoidea were discussed in the previous
section. The phylogenetic relationships of Glyphipterigidae
with Copromorphidae are not readily apparent superficially
but their basic characters indicate a shared origin. Both
Epermeniidae and Glyphipterigidae appear to represent the
more specialized members of the superfamily, although the
Alucitidae are also uniquely divergent.

The glyphipterigid genera of the world, although of
various phenotypes, often show the characteristic feature of
metallic iridescent forewing markings and basic similarities
in the genitalia and wing venation, among other characters.
The genera having larger species with less specialized wing
venation are considered less advanced members of the family. More research is needed on extralimital genera but the
listing in the generic synopsis (p. 34) appears to be relatively
accurate phylogenetically. In the less advanced groups of
genera, from **Chrysocentris** to **Cronicombra**, the genera Cotaena and **Myrsila** are transfers from Sesiidae, **Cotaena** having
been incorrectly transferred to Heliodinidae by Naumann
(1971).

Since so little is known of the biology of glyphipterigids,
the very useful characters of the immature stages are not
usually available for study. The exotic genera, especially

the less advanced genera, are almost entirely unknown biologically. Thus, the phylogeny of Nearctic genera illustrated
in Figure 1 is generalized. **Abrenthia** and **Neomachlotica** are
the least advanced among the Nearctic genera and together
with other less advanced genera, may eventually be grouped
as a subfamily of Glyphipterigidae. However, no clear division at the subfamily level is thus far evident, only a
generalized division occurs between the less advanced and
the more advanced genera. **Drymoana**, new genus, and
Diploschizia are divergent from **Glyphipterix** ancestors.
Drymoana differs principally through wing venation and
unusual labial palpi, while in **Diploschizia** a vein has been
lost in the hind wings. Due to the reduced wing venation
and usually very complex genitalia, I place **Diploschizia** as
the most advanced glyphipterigid. Thus far, it appears that
the genus is restricted to the New World.

To accurately assess the phylogenetic relationships
throughout the family will require further study of the world
genera.

EXCLUDED TAXA

A large number of genera and species have at one time
or another been included in Glyphipterigidae (**sensu lato**). A
catalog (Heppner, 1982b) prepared in conjunction with this
revision lists over 200 generic names (counting synonyms,
emendations, and misspellings) once associated with the
family. This catalog covers the world fauna. Genera formerly placed in the family from the North American fauna
and now excluded are as follows: **Araeolepia** Walsingham
(Plutellidae), **Ellabella** Busck (Copromorphidae), **Hilarographa**
Zeller (Tortricidae), **Homadaula** Zeller (Plutellidae), **Lotisma**
Busck (Copromorphidae), and **Tegeticula** Zeller (Incurvariidae:
Prodoxinae). These genera, as well as one species of "Glyphipterix" belonging to Oecophoridae (now in **Fabiola**), were
discussed more fully and transferred in a recent paper (Heppner, 1978). **Tegeticula** has already been transferred by Davis
(1967).

The remaining North American genera not noted above
now form the Nearctic fauna of the family Choreutidae:
Anthophila Haworth, **Brenthia** Clemens, **Caloreas** Heppner,
Prochoreutis Diakonoff & Heppner, **Choreutis** Hübner,
Hemerophila Hübner, **Tebenna** Billberg, and **Tortyra** Walker.

The Nearctic Choreutidae will be dealt with in a forthcoming revision (Heppner, in press).

TAXONOMIC CHARACTERS

This revision is based predominately on genitalia, wing
venation, and head morphology. The peculiarities and eval-
uation of these and other characters are noted below.

Head

Head structures studied for each species include the
vestiture, labial palpi, eye size, and antenna. Head morphology was studied in detail for the type-species of the
included genera from North America (Figs. 20–29). Head
structures studied for all species are relatively constant in
each genus except **Glyphipterix** where there are long and
short labial palpi, smooth-scaled and rough-scaled labial
palpi, and large and small eyes. **Glyphipterix** is a large
genus but species are known that fill apparent gaps among
the varied species so that the genus does not show any
definable divisions at the generic level.

The labial palpi differentiate between groups having a
long apical segment and those having this segment subequal
or somewhat shorter than the second segment. Only one
species of **Glyphipterix** in North America has very short,
only slightly up-turned labial palpi. Other characters do not
support a division of **Glyphipterix** based on these palpal
differences except at the species-group level.

Maxillary palpi were studied for each genus but were not
used for species differentiation. The haustellum in Glyphipterigidae is well-developed, ranging from relatively short
(**Abrenthia**) to often quite long (**Glyphipterix**), and is always
naked. The antennae are filiform, relatively long, but short
and thickened in some genera (**Abrenthia**). Males always
have a denser setaceous ventral surface on the antennae
than do the females. Ocelli are prominent in all glyphipterigids; they do not appear to provide any characters for
generic or species differentiation. The compound eyes vary
in size between genera and also among species but to a
slighter extent. The ommatidia of the compound eye have
stout setae between the exterior facets (Fig. 48). **Glyphipterix** has the greatest range of eye size among the
species studied.

The anterior tentorial pits are unusually latero-ventrad at the corners of the frons in **Neomachlotica** but the significance of this is not known since this character has not been studied on a comparative basis in Lepidoptera.

Thorax

The coloration and vestiture of the thorax was noted for each species but no other characters are evident that appear useful for generic or species separation. Internal structures, however, are of interest but at the family level. Primarily these include the metafurcal sternum, another character at the family level used by Brock (1971) in his study of ditrysian family relationships. Special dissections are required to study this structure; these were not done since a structure of almost equal value, the type of abdominal articulation, is available on the anterior end of the abdomen routinely mounted on genitalia slides.

Wings

The size of each species was determined by measuring from the forewing base to apex, including the fringe. As noted previously, color is dependent on the extent of diurnal flight under sunlight that a particular specimen was involved in before capture and also, to some extent, the museum age of the specimen. Structural colors, notably the iridescent spots, retain their brilliance in museum specimens as far as has been determined.

The wing venation is relatively similar among the Nearctic genera, with differences occurring principally in the cubital veins of the forewing and the median veins of the hindwing. Overall wing shape also varies among genera, being more broad winged in **Abrenthia** and **Neomachlotica** and related genera, while being more narrow winged in **Drymoana**, new genus, and **Diploschizia**. **Neomachlotica** has the most extensive venational differences, with a very long chorda and strongly convergent cubital veins in the forewing. **Diploschizia** has M3 absent in the hindwings.

Abdomen

The abdomen shows no variations among the genera except in relation to the male genitalia. In **Diploschizia** the ventral splitting of the male posterior abdominal segment is modified to form a large hood over the genitalia. This feature is characteristic of all glyphipterigid males but less developed in other genera. The genitalia are attached by lateral membranes from the tegumen and sometimes also from extensions of the tegumen or transtilla that resemble secondary valvae. Lateral coremata at the base of the last abdominal segment are found in males of some Glyphipterix and one species of Diploschizia.

Male Genitalia

The most typical glyphipterigid male genitalia include a very prominent tuba analis, an ovate ring-like tegumen-vinculum arrangement, a long narrow saccus, simple and setaceous valvae, an incomplete transtilla (as extensions of the dorsal valval bases), a tubular anellus articulated with the valval bases, and an elongate aedeagus with a small tubular cornutus, and a campanulate hood over the ductus ejaculatorius near the aedeagus. Most species of Glyphipterix and most other genera conform to the above genital pattern. Other genera, including Abrenthia, Machlotica, Neomachlotica, and Ussara, have developed coremata on the distal anterior sides of the valvae, as have also a few species of Glyphipterix in the Nearctic; it is not known how extensive this character is in the family throughout the world fauna. Species with valval coremata do not also have coremata on the posterior abdominal segment.

Some species of Glyphipterix have very modified valvae with projections and stout sclerotizations. Species of Diploschizia have the most highly modified male genitalia, often with a complete transtilla and long projections arising from it that resemble secondary valvae, sometimes larger than the actual valvae. These species also have unusual cornuti: long spines in one Nearctic species and possibly deciduous cornuti in other species. The aedeagi of Neomachlotica species are also very modified, with an apical ring of hooks in all known species, but the cornuti are simple tubules.

The extreme genitalia modifications of Diploschizia are considered to represent a more advanced group than Glyphipterix. The modifications of genera like Neomachlotica are difficult to accommodate in relation to other characters that indicate a less advanced position in the family but appear to be separate apomorphic developments among the less advanced genera.

The transtilla appendages of some **Diploschizia** are very similar to such structures found in **Cosmopterix** (Cosmopterigidae) but are apparently of independent derivation.

Female Genitalia

Typical glyphipterigids have a moderately long, simple ovipositor, a small membranous funnel or partly sclerotized cup-like ostium bursae, a short and thin ductus bursae, and a small to moderate ovate corpus bursae. A signum is usually absent but has developed in a number of species in various genera; it does not appear correlated to definable generic limits in itself. An accessory bursa also has developed among various species; it also does not appear to be correlated with other characters. The ostium bursae is sometimes produced on a long sclerotized tubular projection from the edge of the 7th sternite. Females having this feature invariably belong to the same species as males with very long aedeagi (**Diploschizia**). The 8th abdominal sternite is sometimes modified with projections or large plate-like structures but this has only been found in the genus **Diploschizia** and one species of **Glyphipterix**.

Larvae and Pupae

Characters of immature states were not used in the revision below the family level primarily because only two species in the Nearctic have described larvae and pupae, one of which is newly described herein. The available characters of the immature stages were used for higher category classification in comparisons to related families of Ditrysia (Heppner, 1977).

BIOLOGY

The extent of biological information on Glyphipterigidae is so limited and the percentage of species where the host plant is known is so low that little can be conclusively stated regarding general biological properties for the family. Although various authors have often stated that glyphipterigid larvae feed predominately on Gramineae, Cyperaceae, and Juncaceae, this conclusion refers principally to north temperate species of **Glyphipterix** and not to tropical elements of the family. A few north temperate species are

even noted on other plant families in addition to the tropical
species.   Thus, when a greater proportion of the species
become known biologically, host plant associations may prove
to be more extensive among plant families.

It does appear that glyphipterigid larvae are all borers,
whether in seeds, terminal buds, stems, branches or leaves
(in the latter case the larvae are reported to be leaf miners,
a form of borer).   The extent of host specificity is not
known but only 5 species have been reported from more than
one species of host plant.   Co-evolution between glyphipterigids and plant groups has not been documented but several related species of **Glyphipterix** from Europe and North
America have larval hosts in relatively closely related
monocotyledon plant families (Gramineae, Cyperaceae, Araceae, and Juncaceae).

The list at the end of this section includes all known
host plant records for glyphipterigid species of the world.
Out of the 343 described species in the world only 24 have a
host record or probable host association (only 7% of the
world fauna).   For the Nearctic, 4 species have been reared
and 5 species have probable host associations.   The Nearctic
hosts include Cyperaceae (for 3 species), Juncaceae (for 3
species), and Urticaceae (for 3 species).

Most glyphipterigids apparently deposit eggs on plant
surfaces or crevices, inasmuch as the typical ovipositor is
setaceous and relatively unspecialized for piercing.   A few
species have very sharply sclerotized papilla anales edges
and these species undoubtedly pierce the host structure for
egg deposition.   The eggs of **Diploschizia habecki** Heppner
are pale white and rounded.   The eggs of other glyphipterigids are unknown except for the British cocksfoot moth,
Glyphipterix simpliciella (Stephens) (formerly **cramerella** of
authors), which has similar eggs (Chopra, 1925).   Larval
development periods are incompletely known for only three
North American species and are summarized under the respective species discussions (**Neomachlotica spiraea** Heppner,
Diploschizia impigritella (Clemens), and **D. habecki** Heppner.
Chopra (1925) provided a detailed life history of the British
cocksfoot moth mentioned above.   He records 4 larval in-
stars over a period of 4 weeks in England.   A cocoon is spun
away from the host or in a hollowed seed and the winter is
spent in larval diapause.   Pupation occurs the following
spring with adult emergence after about 3 weeks.   He recorded an egg developmental period of 10-15 days for larval
emergence. Stainton (1870) and Empson (1956) also reviewed
the life history of this British moth.

Parasites have been reared from the cocksfoot moth but the scientific names of the 6 hymenopterous species were not noted in Jones and Jones (1964) who reported the parasite information. Additional parasite records are noted at the end of this section.

The larvae boring in seeds typically excavate the last seed they occupy as a pupal chamber. One species, **Diploschizia habecki**, prepares a filigreed area on one side of the host seed case from which the adult escapes. Larvae utilizing the empty seed case have a similar behavior in the British cocksfoot moth but leave only a small hole rather than a filigreed network (Chopra, 1925). Stem borers prepare a pupal chamber close to the stem wall, sometimes near a leaf axil, and leave intact only a thin section of the stem wall through which the adults emerge. A species from Florida, **Neomachlotica spiraea**, often pupates away from the host on protected surfaces in a fluted, filigreed cocoon (Needham, 1955). Pupae are not protruded at ecdysis. Univoltine species in Europe in addition to the British cocksfoot moth have been noted to overwinter as prepupal larvae, with pupation occurring immediately before adult emergence the following spring (Hering, 1932).

Life histories of known species involve mostly one generation per year, predominately among European and northern North American species. Stainton (1870) and Waters (1928) noted that all species in Great Britain are univoltine. The range of collection dates for species distributed in the southeastern United States indicates that these species have multiple or at least bivoltine generations. Several species endemic to Florida, or with a southern intrusion there, seem to have year-round generations. For the more northern, apparently univoltine species in North America, June to July is the most common period of adult activity as far as collection dates indicate, with some species having additional records into August. Species not occurring year-round in the southern regions and in the warmer areas of California, appear to have two or three generations from April to July and August, or even later in Florida. There are two southwestern species and two eastern species that have adult collection records only from September to November. These species are rare in collections but this may be due to the inactivity of most collectors after early September.

Available information indicates that all glyphipterigids are diurnally active as adults. Typically, adults are found in close proximity to their larval host plant or on various flowers. Flower records include **Potentilla** and **Pyracantha** (both in Rosaceae), **Arctostaphylos** (Ericaceae), and **Helianthus, Heliopsis, Pluchea,** and **Solidago** (all in Compositae) (in part, Chambers, 1877b) for North America, while there is a **Prunus** (Rosaceae) (Gozmány, 1954) record for a species in Europe. Empson (1956) refers to numerous flower records for the British cocksfoot moth but unfortunately does not cite them. Adults will often perch on leaves and "twitch" their wings in between short jerking walking movements. In "twitching" their wings they bring the hind legs under the folded wings and push the wings up and out, then move forward a few steps and repeat the wing "twitch." Whether this is a mating display, possibly restricted to males, is not known. These movements have been observed in several species of **Glyphipterix** and **Diploschizia** and the behavior has also been reported by other authors (Chopra, 1925; Empson, 1956; Stainton, 1870). The Florida species of the new genus **Neomachlotica** does not have this wing "twitching" behavior but holds the tips of the forewing in such a way that they curl ventrally when folded over the abdomen in resting position (Needham, 1955).

The Glyphipterigidae are generally of no economic importance but two species are reported as pests of grass crops: the British cocksfoot moth already mentioned previously, **Glyphipterix simpliciella**, and named after the common name for the host, **Dactylis glomerata** Linnaeus (Gramineae), and a New Zealand "cocksfoot moth," **Phryganostola achlyoessa** Meyrick (Ferro, et. al., 1977). The biology of the latter species has not been published in detail.

Known Hosts of Glyphipterigidae

Gramineae

Dactylis glomerata Linnaeus
Glyphipterix fischeriella Zeller [=Glyphipterix simpliciella (Stevens)], seed borer (Chopra, 1925; Hartman, 1880; Stainton, 1870), Europe.
Festuca sp. (?)
Glyphipterix fuscoviridella (Haworth), culm borer? (Hering, 1957), England.
Pleioblastus variegatus Makino var. viridis Makino
Glyphipterix semiflavana Issiki, culm borer (Moriuti, 1960), Japan.

Cyperaceae

Carex vulpina Linnaeus
  Glyphipterix forsterella (Scopoli), seed borer (Meyrick, 1928), England.
Cyperus esculentus Linnaeus
  Diploschizia impigritella (Clemens), stem and leaf axil borer (Frick, pers. comm.), Mississippi.
Cyperus rotundus Linnaeus
  Diploschizia impigritella (Clemens), ibid.
Eriophorum angustifolium Roth
  Glyphipterix haworthana (Stephens), seed borer (Stainton, 1897), Europe.
Eriophorum gracile Koch
  Glyphipterix haworthana (Stephens), seed borer (Hartmann, 1880), Europe.
Eriophorum lanatum [unknown, probably **E. latifolium** Hoppe]
  Glyphipterix haworthana (Stephens), seed borer (Stainton, 1855), Europe.
Gahnia setifolia Streud
  Glyphipterix calliactis Meyrick, stem pith borer (Hudson, 1928), New Zealand.
  Glyphipterix leptosema Meyrick, outer stem borer (Hudson, 1928), New Zealand.
Rhynchospora corniculata (Lamarck) Gray
  Diploschizia habecki Heppner, seed borer (Heppner, 1981a), Florida and Georgia.
Schoenus nigricans Linnaeus
  Glyphipterix schoenicolella (Boyd), seed borer (Hartmann, 1880; Stainton, 1870), Europe.

Araceae

Acorus gramineus Soland
  Lepidotarphius perornatella Walker, stem borer (Kodama, 1961), Japan.

Juncaceae

Juncus glomeratus Thunberg [= **J. effusus** Linnaeus]
  Glyphipterix thrasonella (Scopoli), stem borer (Hartman, 1880), Europe.
Juncus sp.

Glyphipterix iocheaera Meyrick, stem borer (Hudson, 1928), New Zealand.
Glyphipterix juncivora Heppner, (present data), Rockies and Great Basin
Glyphipterix montisella Chambers, ibid.
Glyphipterix roenastes Heppner, ibid.
Pantosperma holochalca Meyrick, (Hudson, 1928), New Zealand.
Luzula campestris De Candolle
    Glyphipterix fuscoviridella (Haworth), stem borer (Meyrick, 1928), Europe.
Luzula pilosa Willdenow [=**L. vernalis** De Candolle]
    Glyphipterix bergstraesserella (Fabricius), stem borer (Fettig, 1882), Europe.
Luzula sp.
    Glyphipterix bergstraesserella (Fabricius), stem borer (Spuler, 1910), Europe.

Piperaceae

Piper aduncus Linnaeus
    Ussara eurythmiella Busck, leaf feeder (Busck, [1934]), Cuba (the Cuban **Ussara** species may not be conspecific with the named species from Panama).
Piper auritum Humbolt, Bonpland & Kunth
    Ussara eurythmiella Busck, ibid.

Urticaceae

Boehmeria cylindrica (Linnaeus) Schwarz
    Neomachlotica spiraea Heppner, bud and stem borer (Needham, 1955), Florida.
Urtica sp.
    Glyphipterix powelli Heppner, (present data), California.
    Glyphipterix urticae Heppner, (present data), Rockies and Great Basin.

Crassulaceae

Cotyledon umbilicus Linnaeus (cited as **Umbilicus pendulinus** De Candolle)
    Glyphipterix umbilici Hering, leaf miner (Hering, 1927), Europe.
Securigera sp.

Glyphipterix equitella (Scopoli), on leaves (Hering,
    1957), Europe.
Sedum acre Linnaeus
  Glyphipterix equitella (Scopoli), leaves and stem
    (Hartmann, 1880; Stainton, 1870), Europe.
Sedum album Linnaeus
  Glyphipterix equitella (Scopoli), ibid.
Sedum sexangulare Linnaeus
  Glyphipterix equitella (Scopoli), ibid.

Known Parasites of Glyphipterigidae

Hymenoptera: Ichneumonidae

Angitia glabricula Holmgren [=**Horogenes glabricula**
    (Holmgren)]
  Glyphipterix forsterella (Fabricius), (Schütze & Ro-
    man, 1931; Thompson, 1946), Europe.
Campoplex tumidulus Gravenhorst
  Glyphipterix haworthana (Stephens), (Morley & Rait-
    Smith, 1933; Thompson, 1946), Europe.
Pimpla nucum Ratzeburg
  Glyphipterix haworthana (Stephens), ibid.

GEOGRAPHICAL DISTRIBUTION

World Fauna

The world fauna of Glyphipterigidae is summarized by
genera in the generic synopsis on page 34. The majority of
the world species are in one widespread genus, **Glyphipterix**,
with 276 described species, including the 21 new species
added to the genus herein.   A large proportion of **Glyphipterix** species are Neotropical and from the Old World
tropics, almost entirely described by Meyrick.   Another large
number of species are Palearctic.   Exclusive of northern
Mexico, the Nearctic fauna of **Glyphipterix** comprises 29
species, including the 21 new species.   The remaining 25
genera (including the new genus described herein) are virtually all Pantropical and many of them are monobasic.   As
various areas of the world become better known, the number
of monobasic genera will undoubtedly be reduced by the
addition of other species.

Centers of greatest diversity apparently include the northern Neotropical and the Oriental regions. Judging from the accumulated undescribed species from the Neotropical region, this area will have a large percentage increase in species numbers. Unusual endemic genera are found in Pantropical regions and in New Zealand. South Asia has species of apparently early origin, with some relationship to an unusual species of **Glyphipterix** from North America. Both the Palearctic and Nearctic regions have large numbers of closely related species in **Glyphipterix**.

Nearctic Fauna

The northern Nearctic region, exclusive of Nearctic portions of Mexico, comprises a fauna of 36 species, primarily in **Glyphipterix** (27 species). The greatest faunal relationship is Palearctic, with one Holarctic species. There is also a significant Neotropical element in the fauna, composed primarily of **Abrenthia** (1 species), **Neomachlotica** (1 species), and **Diploschizia** (6 species). Each of these genera has Neotropical members except the endemic **Abrenthia** which, however, is closely related to Neotropical genera. A second endemic Nearctic genus, **Drymoana**, new genus (1 species), has obscure relationships but perhaps is closely related to Holarctic elements of **Glyphipterix**.

North America north of Mexico has three main regions of species concentrations. The western region, from the Rockies to the Pacific Coast, is the area of greatest diversity and endemism. California has the highest number of endemic species (7 species), followed by Arizona (6 species) and, in the East, by Florida (5 species). The western **Glyphipterix** include several species that appear either very plesiomorphic (1 species in Wyoming), with nearest relatives possibly in south Asia, or endemic (**bifasciata** group) and relatively isolated in the genus. Western North America includes species only of **Glyphipterix** and one widespread species of **Diploschizia**. The greatest species radiation has occurred in the western region, with 3 species-groups (two having 4 species each and one having 6 species) of **Glyphipterix**.

The northeastern quarter of the continent is another region with species not found elsewhere (3 species) but it also has several widespread species (4 species). The fauna is much less diverse than in the western half of North America but the area is relatively uncollected for these diurnal microlepidoptera. Several of the eastern species are poorly represented in collections and for 3 species only one sex is known. Further collecting may increase the species number of the northeastern fauna.

The Southeast includes all the range extensions of tropical genera or relatives, except one widespread species in **Diploschizia** which ranges north and west. Florida has the greatest number of endemic species in this region (5 species), with a primarily tropical fauna. Two species have a southeastern distribution, from North Carolina to central Texas, although further collecting may increase the number of species having this range. Subtropical Texas, near Brownsville, has no recorded species of Glyphipterigidae; thus, further collecting there may furnish additions to the North American fauna from northern Mexico or as endemics.

Only one species, **Diploschizia impigritella** (Clemens), is widespread over North America; it is distributed in a northerly arch from the East to the Pacific Coast, generally bypassing the Great Plains. Two other species are transcontinental across southern Canada or nearly so, ranging south into the United States along each coast: **Glyphipterix haworthana** (Stephens) and **Glyphipterix sistes,** new species. One species, **Glyphipterix bifasciata** Walsingham, is quite widespread along the Pacific Coast, with limited records from the northern Great Basin, ranging north along the coast to southern Alaska.

Distributional patterns for Nearctic Glyphipterigidae other than the localized endemics involve seven patterns: 1) Californian, primarily Coast Range, southern Cascades, or Sierran; 2) Pacific Coast, ranging from northern California to southern Alaska; 3) Great Basin and southern Rockies, including the east slope of the Sierra Nevada of California; 4) northeastern, from Nova Scotia to New England but ranging west to the Yukon; 5) eastern, from Nova Scotia to Manitoba and Kansas, often south to Florida; 6) southeastern, from North Carolina to Texas; and 7) Floridian, mainly southern subtropical Florida.

As noted earlier, the North American fauna of microlepidoptera is still relatively poorly known and in Glyphipterigidae several additional new species would be expected with further collecting. Of the 36 species discussed herein, 3 species are known only from females, 5 species are known only from males, and 16 species are represented by only one or a few specimens of one of the sexes or both sexes. A few species have long series of males but only one or two females and 9 species are known only from the type locality.

Generic Synopsis of Glyphipterigidae[2]

*Chrysocentris Meyrick, 1914	Ethiopian, Oriental (8 sp.)
*Irinympha Meyrick, 1932	Ethiopian (1 sp.)
*Ernolytis Meyrick, 1922	Oceania (1 sp.)
*Carmentina Meyrick, 1930	Oriental, Australian (6 sp.)
*Cotaena Walker, [1865]	Neotropical (2 sp.)
*Myrsila Boisduval, [1875]	Neotropical (1 sp.)
*Lepidotarphius Pryer, 1877	Oriental, east Palearctic (1 sp.)
*Tetracmanthes Meyrick, 1924	Ethiopian (1 sp.)
*Phalerarcha Meyrick, 1913	Neotropical (2 sp.)
*Cronicombra Meyrick, 1920	Neotropical (6 sp.)
*Taeniostolella Fletcher, 1940	Neotropical (2 sp.)
*Machlotica Meyrick, 1909	Neotropical (2 sp.)
Abrenthia Busck, 1914	Nearctic (1 sp.)
Neomachlotica Heppner, 1981	Nearctic, Neotropical (4 sp.)
*Trapeziophora Walsingham, 1892	Neotropical (1 sp.)
*Rhabdocrates Meyrick, 1931	Neotropical (1 sp.)
*Ussara Walker, 1864	Neotropical, Oriental, Ethiopian (13 sp.)
*Sericostola Meyrick, 1927	Neotropical (1 sp.)
*Electrographa Meyrick, 1912	Oriental (1 sp.)
*Apistomorpha Meyrick, 1881	Australian (1 sp.)
*Phryganostola Meyrick, 1881	Australian (6 sp.)
*Pantosperma Meyrick, 1888	Australian (1 sp.)
*Circica Meyrick, 1888	Australian (2 sp.)
Drymoana Heppner, new genus	Nearctic (1 sp.)
Glyphipterix Hübner, [1825]	World (276 sp.)
Diploschizia Heppner, 1981	Nearctic, Neotropical (9 sp.)

---

[2] The asterisk (*) indicates extralimital genera.

Nearctic Checklist

Abrenthia Busck, 1915
1. **cuprea** Busck, 1915   New York to Florida,
                                             west to Illinois

Neomachlotica Heppner, 1981
2. **spiraea** Heppner, 1981   Florida

Drymoana Heppner, n.gen.
3. **blanchardi** Heppner, n.sp.   North Carolina to
                                             Florida; Texas

Glyphipterix Hübner, [1825]
    Heribeia Stephens, 1829
    Aechmia Treitschke, 1833
    Aecimia Boisduval, 1836, missp.
    Glyphipteryx (Curtis, 1827) Zeller, 1839, emend.
    Glyphiteryx Fischer von Röslerstamm, 1841, missp.
    Anacampsoides Bruand d'Uzelle, 1850, nom. oblit.
    Glypipteryx Stainton, 1854, missp.
    Glyphopteryx Herrich-Schäffer, 1854, missp.
    Glyphiptoryx Mann & Rogenhofer, 1878, missp.
    Glyphptieryx Turati, 1879, missp.
    Glyphipterys Christoph, 1882, missp.
    Glyphyteryx Hampson, 1918, missp.
    Glyphteryx Watt, 1920, missp.

The **brauni** species-group

4. **brauni** Heppner, n.sp.   Wyoming

The **circumscriptella** species-group

5. **circumscriptella** Chambers, 1881
    a) **circumscriptella** Chambers, 1881 Massachusetts to
                                             Illinois; Texas
         circumstripta Dyar, 1900, missp.
    b) **apacheana** Heppner, n.subsp.   Arizona

The **quadragintapunctata** species-group

6. **quadragintapunctata** Dyar, 1900   Ohio to Kansas

The **powelli** species-group

7. **powelli** Heppner, n.sp.
    a) **powelli** Heppner, n.subsp.   California
    b) **jucunda** Heppner, n.subsp.   Washington
8. **urticae** Heppner, n.sp.
    a) **urticae** Heppner, n.subsp.   Colorado to New
                                            Mexico, Utah,
                                            California

    b) **sylviborealis** Heppner, n.subsp. Manitoba to Alberta

The **bifasciata** species-group

9. **bifasciata** Walsingham, 1881   Montana to British
                                            Columbia, to
                                            California; Utah,
                                            Nevada

10. **hypenantia** Heppner, n.sp.   California
11. **yosemitella** Heppner, n.sp.   California
12. **unifasciata** Walsingham, 1881   California

The **haworthana** species-group

13. **haworthana** (Stephens, 1834)   Nova Scotia to
                                            Manitoba, to
                                            Northwest Terr.;
                                            Europe
        haworthella (Stephens, 1829), nom.nud.
        zonella (Zetterstedt, [1839])
        howarthana Jordan, 1886, missp.
14. **sistes** Heppner, n.sp.
    a) **sistes** Heppner, n.subsp.   Alaska to California
    b) **viridimontis** Heppner, n.subsp.   Nova Scotia to
                                            Vermont

The **californiae** species group

15. **californiae** Walsingham, 1881   Oregon to California
16. **feniseca** Heppner, n.sp.   California
17. **juncivora** Heppner, n.sp.   Alberta to New
                                            Mexico, Utah to
                                            Arizona
18. **sierranevadae** Heppner, n.sp.   California
19. **arizonensis** Heppner, n.sp.   Arizona

20. **roenastes** Heppner, n.sp.   Colorado to new
                                          Mexico, Utah to
                                          Arizona

The **montisella** species-group

21. **chiricahuae** Heppner, n.sp.   Arizona
22. **hodgesi** Heppner, n.sp.   Arizona
23. **saurodonta** Meyrick, 1913   Ontario to West
                                          Virginia
24. **cherokee** Heppner, n.sp.   Tennessee
25. **chambersi** Heppner, n.sp.   Manitoba to
                                          Kentucky
26. **montisella** Chambers, 1875   Montana to New
                                          Mexico, Utah to
                                          Arizona; California
      montinella Chambers, 1877, missp.
      montella Meyrick, 1913, emend.
27. **flavimaculata** Heppner, n.sp.   California
28. **melanoscirta** Heppner, n.sp.   Arizona
29. **santaritae** Heppner, n.sp.   Arizona

The **ruidosensis** species-group

30. **ruidosensis** Heppner, n.sp.   New Mexico

Diploschizia Heppner, 1981

31. **lanista** (Meyrick, 1918)   North Carolina to
                                          Florida; Louisiana
32. **minimella** Heppner, 1981   Florida
33. **habecki** Heppner, 1981   Georgia to Florida
34. **regia** Heppner, 1981   Florida
35. **impigritella** (Clemens, 1863)   Newfoundland to
                                          Florida, Manitoba
                                          to Texas; British
                                          Columbia to
                                          California; Nevada
      exoptatella (Chambers, 1875)
36. **kimballi** Heppner, 1981   Florida

SYSTEMATIC TREATMENT

Glyphipterigidae
(Figs. 128, 130-132)

Type-genus: **Glyphipterix** Hübner, [1825]
    Glyphipterygidae Stainton, 1854a:103 (type-genus
    Glyphipteryx (Curtis, 1827) Zeller, 1839, invalid
    emendation (=**Glyphipterix** Hübner, [1825]), not
    sensu Curtis, 1827 (Agonoxenidae)); Stainton,
    1854b:169; 1859a:153; 1859b:362; Walker, 1864:
    837; Stainton, 1867:85; Heinemann, 1870:392;
    Wocke, 1871:309; Morris, 1872:133; Bang-Haas,
    1875:35; Millière, [1876]:347;Wocke, [1876]:393;
    Hartmann, 1880:93; Meyrick, 1880:205; Donckier
    de Donceel, 1882:138; Fettig, 1882:166; Curo,
    1883:68; Jourdheuille, 1883:190; Moore, [1887]:
    524; Riley, 1891:104; Anonymous, 1897:2551;
    Rebel, 1901:129; Schütze, 1902:25; Turner, 1903:
    76; Crombrugghe de Picquerdaele, 1906:1; Nick-
    erl, 1908:1; Spuler, 1910:296; Meyrick, 1913b:23;
    1914b:284; 1914c:1; Martini, 1916:132; Rebel,
    1916:157; Barnes & McDunnough, 1917:181; Jan-
    se, 1917:203; Meyrick, [1920]:1003; 1921b:184;
    Forbes, 1923:350; Braun, 1924:244; Comstock,
    1924:633; Chopra, 1925:359; Handlirsch, 1925:878;
    Janse, 1925:330; Tillyard, 1926:422; Lycklama,
    1927:13; Philpott, 1927:337; Hudson, 1928:305;
    Leonard, 1928:554; Meyrick, 1928: 705; Vorbrodt,
    1928:115; Matsumura, 1931:1078; Vorbrodt, 1931:
    122; Hering, 1932: 175; Eckstein, 1933:108: Ster-
    neck & Zimmermann, 1933:78; Meyrick, 1935:85;
    Pierce & Metcalfe, 1935:40; Amsel, 1936:353;
    Möbius, 1936:133; Le Marchand, 1937a:189;
    1937b:217; Fletcher, 1938:110; Wu, 1938:378;
    Börner, 1939:1423; Hudson, 1939:456; McDun-
    nough, 1939: 83; Naumann, 1939:119; Comstock,
    1940:633; Fletcher & Clutterbuck, [1914]:104;
    Jäckh, 1942:194; Costa Lima, 1945:308; Heslop,
    1945:27; Kloet & Hincks, 1945:132; Le Marchand,
    1945:100; Procter, 1946:318; Turner, 1947:321;
    Viette, 1947:40; Imms, 1948:449; Lhomme, 1948:

493; Viette, 1949; 21; Amsel, 1950:26; Diakonoff,
1950:175: Bourgogne, 1951:386; Obenberger, 1952:
28; Börner, 1953:397; Ford, 1954:95; Wörz, 1954:
83; Clarke, 1955:25; Toll, 1956:3; Klimesch, 1961:
723; Kodama, 1961:35; Bleszynski, et al., 1965:
413; Brock, [1968]:245; Diakonoff, [1968]: 188;
Klimesch, 1968:149; Brock, 1971:80; Duckworth,
1971:1; Chinery, 1972:185; MacKay, 1972:19;
Davies, 1973:206; Hannemann & Urban, 1974:320;
Dugdale, 1975:579; Diakonoff, 1977a:76; 1977b:
171; 1977c:3; Diakonoff & Heppner, 1977:81;
Kuznetsov & Stekolnikov, 1977:27; Moriuti, 1977:
15; Diakonoff, 1978:45.

Aechmidae Bruand d'Uzelle, 1850:48 (type-genus:
Aechmia Treitschke, 1833 (=**Glyphipterix** Hübner,
[1825]).

Gliphipterygidae Chambers, 1880b:199, missp.

Glyphipterigidae.— Rosenstock, 1885:438; Ford,
1949:126; Inoue, 1954:48; Turner, 1955:163; Inoue,
et al., 1959:273; Common, 1966:42; Bradley, &
Pelham-Clinton, 1967:126; Clarke, 1969:56; Common, 1970a:233; 1970b:810; Alford, 1971:172;
Krogerus, et al., 1971:17; Common, 1974:104;
Fibiger & Kristensen, 1974:9; Common, 1975:198;
Watson & Whalley, 1975:210; Diakonoff, 1976:82;
Diakonoff & Arita, 1976:179; Karsholt & Nielson,
1976:24; Duckworth & Eichlin, 1977:5; Heppner,
1977:124; Richards & Davies, 1977:1111; Heppner, 1978:48; Diakonoff, 1979:291; Leraut,
1980:81; Heppner, 1981a:309; 1981b:479; Heppner
& Duckworth, 1981:1; Heppner, 1982a:38; 1982b:
220; 1982c:704; 1982d:257; 1983a:25; 1983b:99;
1984:54; Covell, 1984:432; Kyrki, 1984:78.

Glyphipserygidae Hartig, 1956:125, missp.

Glyphiptreygidae Hruby, 1964:259, missp.

Adults small to moderate (2-15 mm. forewing length).
Head: frons and vertex smooth-scaled; labial palpus usually
long and upturned, dorso-ventrally flattened on apical
segment, rarely somewhat short and not distinctly dorso-
ventrally flattened; labial palpus usually with apical segment
longest or subequal to 2nd segment; venter of segment 2
usually smooth-scaled, sometimes roughened, rarely with long
scale tuft; maxillary palpus 2- to 4-segmented; haustellum
naked, developed mandibles; moderate; pilifers large or small;
ocellus well-developed, prominent; eye usually large; antenna

usually moderate, thin, filiform, sometimes short and thickened, males with longer ventral setae than females. **Thorax:**
smooth-scaled; relatively normal in relation to wing size.
Forewing: usually elongate-oblong, sometimes shortened and
somewhat broad; apex usually acute-rounded but sometimes
more broadly rounded; margins usually somewhat convex except in very elongate-winged genera; termen with rounded
falcate indentation from apex; tornus merges with termen;
anal angle rounded; pterostigma well-developed; chorda
usually present or vestigial, sometimes very long; cell usually
without median vein or with vestigial vein; radial veins to
costal margin except R5 usually to termen or apex; M1-M3
usually parallel to termen, free; cubital veins often parallel
to tornus, rarely with CuA2 broadly convergent to CuA1 at
tornus; CuA2 near end of cell; CuP present at margin, extended as fold; A1+2 with short to moderate basal fork; A3
very reduced at anal margin; A4 absent. **Hindwing:** elongate-
oblong or relatively narrow with acute pointed apex; costal
margin relatively straight but somewhat sinuate; dorsal
margin straight to rounded or abrupt anal angle; apex
rounded or acute; termen distinct to rounded tornus or
indistinct; Rs to or before apex, rarely vestigial near wing
base, usually free of M1 but sometimes meeting M1 at end
of cell; M1 and M2 usually closer than M2 to M3; M3 rarely
absent; M3 often meeting CuA1 at end of cell; crossvein
usually oblique toward tornus, rarely directed toward dorsal
margin; CuA1 and CuA2 usually parallel; CuA2 from near
end of cell; CuP present at margin, extended as fold; A1+2
with long or short basal fork; A3 long, from base of A2 fork;
A4 at anal margin. **Abdomen:** elongate; coremata rarely present on last male segment latero-ventrally; articulation with
thorax (Fig. 128) of tineoid type; posterior end rarely
strongly modified as ventrally split hood for male genitalia
(Figs. 130-131). **Male genitalia:** uncus and gnathos absent;
tuba analis well-developed, often with lateral sclerotized
ridges; socius absent; tegumen often narrow but stout, merged to vinculum to form elongate ring-like structure; tegumen rarely with lateral appendage attached distally to male
abdominal modification as , hood, appearing as secondary
valvae (pleural lobes); vinculum convex, subquadrate or
rectangular; saccus usually well-developed, narrow, often
long, sometimes very broad, or reduced or absent; valva
usually simple, setaceous, long or short, sometimes variously
modified and stoutly sclerotized, with saccular point or other
projected spines; coremate setae rarely on anterior side of
valva; anellus (Fig. 132) tubular, with ventral split in

sclerotized portion but there membranous to form tube, long or short, apically setaceous or without setae; anellus base articulated with valval base or fused to valval base; transtilla present and fused to valval base or only as extensions of valval base, rarely with long setaceous appendages appearing as secondary valvae; aedeagus attached to distal end of anellus, usually elongate, sometimes very long, with small phallobase, with apical spicule collar or smooth, rarely modified with ring of apical hooks; cornutus usually a small tubule, rarely absent or possibly deciduous, or as 3 large spines; vesica usually without spicules; ductus ejaculatorius usually with large campanulate hood near aedeagus or a hood-like structure distant from aedeagus. **Female genitalia:** ovipositor short or long, usually unmodified, rarely with projections on 8th sternite; papilla analis usually simple, setaceous, rarely sclerotized with sharp edges; apophyses short or long, usually thin; ostium bursae usually a membranous funnel often with an anterior sclerotized cup or mostly sclerotized, on intersegmental membrane between sternites 7 and 8, rarely modified as a central cone in a sunken circular sterigma or as an extended sclerotized narrow tube; ductus bursae thin, often sclerotized or membranous and wider; ductus seminalis from ductus bursae near ostium or near bursa or from special sclerotized junction; bulla seminalis small; corpus bursae usually ovate, moderate or elongate-ovate, sometimes small; accessory bursa sometimes present from anterior end of bursa; signum usually absent, sometimes present as spicule patch or fused spicule area or line, or row of large fused teeth-like spines.

Larva.- Generally unknown but known genera with head having 2 adfrontal setae (1 in **Glyphipterix**?); stemmata in semi-circle; 2 setae in L-group of prothorax; prolegs on segments 3-6 vestigial; crochets secondarily absent, rarely present in uniordinal lateral penellipse; spiracle often projected on a cone-like structure, especially on prothorax and 8th abdominal segment; tergite 10 sometimes as large sclerotized plate with large spines posteriorly.

Pupa.- Unspined dorsally but with 2 setae on each segment; head with sharp horn-like projections; maxillary palpus small; cremaster as hook-tipped spines on venter of last segment; spiracles on short cylindrical projections.

Key to North American Genera

1.   Hindwing vein M3 absent (Fig. 57) ....
        .... Diploschizia (p. 133)
        Hindwing vein M3 present 2

2(1).   Forewing veins CuA1 and CuA2 strongly convergent
        at termen (Fig. 53) ....
        .... Neomachlotica (p. 45)
        Forewing veins CuA1 and CuA2 parallel or divergent
        at termen 3

3(2).   Labial palpi with very long scale tuft ventrally,
        longer than 2nd and basal segments combined
        (Fig. 32) .... Drymoana (p. 49)

        Labial palpi smooth or with moderate tufts not
        longer than 2nd and basal segments combined ...4

4(3).   Forewing vein R5 curved up to apex (Fig. 52)............
        .... Abrenthia (p. 42)
        Forewing vein R5 curved down away from apex,
        even if stalked with vein R4 (Figs. 55-56) ....
        ....Glyphipterix (p. 53)

Abrenthia Busck
(Figs. 20-21, 30, 36, 42, 52)

Abrenthia Busck, 1915:87 (type-species: Abrenthia
cuprea Busck, 1915, original designation); Barnes
& McDunnough, 1917:81; Forbes, 1923: 352;
Fletcher, 1929:1; McDunnough, 1939:83; Heppner,
1981b:481, 1982a:45, 1982b: 236; 1983a:25.

Among North American genera of Glyphipterigidae,
Abrenthia is similar to **Neomachlotica**, from which it is
readily distinguished by the parallel CuA1 and CuA2 of the
forewings.

Adults small (4.0-5.5 mm. forewing length). Head (Figs.
20, 30): frons and vertex smooth-scaled; labial palpus recurved, basal and 2nd segments subequal, apical segment half
as long, segments smooth-scaled; maxillary palpus (Fig. 21)
3-segmented; haustellum (Fig. 36) developed; pilifers large;
ocellus moderate; antenna (Fig. 42) relatively thick, short.
Thorax: smooth-scaled. **Forewing** (Fig. 52): elongate, with

long pterostigma; costal margin straight to pterostigma, then rounded to abrupt apex; termen rounded evenly through tornus to convex dorsal margin; chorda developed; no vein in cell; Sc to costal margin; R3 and R4 approximate at end of cell; M1–M3 evenly spaced at end of cell; M3 distant from CuA1 at end of cell; CuA1 and CuA2 parallel; CuP present at margin; A1+2 short-stalked at base; A3 short. **Hindwing:** elongate; costal margin sinuate; apex rounded; termen oblique to rounded tornus and long straight dorsal margin; anal angle rounded; Sc+R1 to near apex; Rs to apex, separated from base of M1; M1 and M2 divergent; M2 distant from M3; M3 short-stalked with CuA1; CuA1 and CuA2 slightly divergent; A1+2 with moderate basal stalk; A3 long; A4 at anal margin. **Abdomen:** posterior segment modified in male as genitalia hood with ventral split; no coremata. **Male genitalia:** tegumen stout, protruded to point dorsally; vinculum broad, with narrow saccus; tuba analis prominent and long; valva elongate, simple but with latero–distal corematal setae from near base, mesal side setaceous; valval base prolonged as base of tubular anellus; anellus a lightly sclerotized tube, with aedeagus attached at tip; aedeagus with phallobase and setaceous apical band; cornutus a tubule. **Female genitalia:** ovipositor with sharp papilla analis; papilla analis elongate, somewhat sclerotized; apophyses subequal, moderate; ostium membranous, with indistinct opening on intersegmental membrane between segments 7 and 8; ductus bursae very thin, short, gradually widening to bursa; corpus bursae ovate, simple; signum absent; ductus seminalis arising midway from ostium; bulla seminalis small. **Larva:** unknown. **Pupa:** unknown.

Remarks.- **Abrenthia** is monobasic and presently known only from North America. It is not a junior synonym of **Machlotica**, as though by Meyrick (1913b, 1914c), nor with any other Neotropical genus. It is most closely related to the Neotropical **Neomachlotica** and **Trapeziophora**, differing principally in wing venation.

No biological information is known for **Abrenthia**.

Abrenthia cuprea Busck
(Figs. 2, 58–59, 133–134, 203–204)

Abrenthia cuprea Busck, 1915:87; Barnes & McDunnough, 1917:81; Forbes, 1923:352; Leonard, 1928: 554; McDunnough, 1939:83; Heppner, 1982a:45, 1982b:236; 1983a:25.

The purplish, indistinctly marked forewings and usually bronze-colored patagia and head easily distinguish this species from any other New World glyphipterigid.

Male (Fig. 58). - 4.0-5.5 mm. forewing length. **Head:** bronze iridescent (rarely fuscous), sometimes becoming buff on frons; labial palpus fuscous distally with white-buff on dorsal surface and white ventrally from middle of segment 2 to base; antenna with fuscous scales dorsally. **Thorax:** fuscous, some bronze iridescence; patagia bronze iridescent; ventrally white-buff; legs fuscous with white at segmental joints. **Forewing:** uniformly fuscous with about 12 single-scale longitudinal striae of silver iridescence, giving an overall purplish appearance at certain light angles; striae with green to blue iridescence at different angles of view; short silver fascia at pterostigma; termen from pterostigma to beyond tornus with line of silver iridescence broken into several spots; fringe dark fuscous; venter uniformly gray fuscous. **Hindwing:** fuscous; fringe white with basal line of fuscous; venter gray fuscous. **Abdomen:** fuscous with silvery scale row on posterior of each segment; venter mostly white. **Genitalia** (fig. 133): tuba analis long, distinct; tegumen greatly elongated to dorsal point; vinculum convex, gradually merging to narrow saccus of equal length as vinculum; valva elongate, oblong with rounded apex, finely setaceous; valva with corematal setae from near base on anterior side; anellus tubular, elongate, without setae; aedeagus (Fig. 134) very elongate and narrow, 3/4 length of valva, with short phallobase, long tubule cornutus, and spiculed vesica; ductus ejaculatorius with campanulate hood near aedeagus (4 preparations examined).

Female (Fig. 59) - 5.4 mm. forewing length. Similar to male. **Genitalia** (Fig. 204): ovipositor relatively short; papilla analis as long sclerotized pleurite, setaceous laterally; apophyses well-formed, sub-equal; ostium bursae (Fig. 203) small, cup-shaped, membranous, merging into ductus bursae via a short narrow funnel; ductus bursae membranous, very thin to ductus seminalis opening near mid-point, then gradually enlarging to bursa; corpus bursae ovate, membranous, without signum. (1 preparation examined).

Types.- Lectotype male, by present designation: Roxboro, [Philadelphia Co.], Pennsylvania, 21 Jun, F. Haimbach (USNM), (Labelled additionally: "Type, 19239 USNM" and "Lectotype male, **Abrenthia cuprea** Busck, by Heppner, '76."). Paralectotypes (2 males): **Pennsylvania.**- same locality as

lectotype, 20 Jun (1 male), F. Haimbach (USNM) ("Cotype").
Virginia.- Fairfax Co.: Falls Church, 20 Jul 1914 (1 male), C.
Heinrich (USNM) ("Cotype").

Additional specimens (5 males, 1 female).- **Florida.-**
Alachua Co.: Univ. Fla. Hort[iculture] Unit, 9 mi. NW.
Gainesville, 24/25 Mar 1975 (1 male), ex Malaise trap, G. B.
Fairchild (JBH). Highlands Co.: Archbold Biol. Sta., 22 Aug
1978 (1 male), H. V. Weems, Jr. and S. Halkin (FSCA). **Illi-
nois.-** Putnam Co.: 18 Jun 1939 (1 male), M. O. Glenn
(USNM). **Maryland.-** Montgomery Co.: Plummers Is., 8 Jul
1968 (1 female), P. J. Spangler (USNM). **New York.-** Erie
Co.: Protection, 15 Jun 1918 (1 male), W. Wild (CU). **Ohio.-**
Hamilton Co., Cincinnati, 10 Jun 1927 (1 male), A. F. Braun
(ANSP).

Distribution.- (Fig. 2). New York to Florida, west to
Illinois.

Flight period.- March (Florida); June to July (northern
states).

Hosts.- Unknown.

Biology.- Unknown. Specimens thus far have been collected from the eastern deciduous forest habitat.

Remarks.- The single male from Florida has an overall
blue iridescence to the forewing, head and patagia, as compared to the purple of the forewings and bronze of the patagia and head of northern specimens. The genitalia, however, are identical and it appears best at this time to consider the Florida population only as a color form, perhaps
gradually merging with the northern form from Florida to
Virginia. In **Glyphipterix** it is often found that the male
genitalia will be virtually identical among closely related
species while the female will show specific differences; thus,
a female must be studied from Florida to make any further
conclusions about this population.

Neomachlotica Heppner, 1981
(Figs. 22-23, 31, 37, 43, 48, 53)

Neomachlotica Heppner, 1981b:479. (type-species: **Neomachlotica spiraea** Heppner, 1981, original designation);
Heppner, 1982a:45, 1928b:261; 1984:54.

This genus is distinguished from related genera by the
convergent CuA1 and CuA2 at the termen of the forewing.

Adults small (3.2–4.0 mm. forewing length). Head (Figs. 22, 31): frons and vertex smooth-scaled; labial palpus recurved and very dorsoventrally flattened on apical 2 segments, with basal and 2nd segments subequal in length, apical segment twice as long as basal segment; maxillary palpus (Fig. 23) 3-segmented with very long 2nd segment; haustellum (Fig. 37) developed; pilifers large; eye moderate; ocellus moderate (Fig. 48); antenna (Fig. 43) normal. Thorax: smooth-scaled. Forewing (Fig. 53): oblong, with pterostigma; costal margin straight to pterostigmal convexity, then rounded to rounded apex; termen very oblique to indistinct tornus; dorsal margin straight to rounded anal angle; chorda developed, with central vertical vein; no vein in cell; Sc to costal margin before half of wing distance; R1-R5 to costal margin; M1 to apex; M1-M3 equidistant at end of cell; CuA2 distant from end of cell and greatly convergent to CuA1 at termen; CuP present at tornus; A1+2 with moderate basal stalk; A3 short. Hindwing: with Sc+R1 to 3/4; Rs directed up to costal margin before apex; M1 and M2 close together at end of cell, distant from M3: M3 approximate to CuA1 at end of cell; CuA2 nearly parallel to CuA1; A1+2 with long basal stalk; A3 short; A4 vestigial. Abdomen: posterior segment modified in males as genitalia hood with ventral split; no coremata. Male genitalia: tegumen stout and fused with broad vinculum; saccus absent; tuba analis prominent, valva simple, setaceous mesally, with large coremata on distal side near base; valval base formed into elongate dorsal transtilla process and ventral process for base of anellus; anellus a short tube with aedeagus attached at tip; aedeagus without phallobase, with enlarged tip having a ring of recurved hooks and a band of spines and setae. Female genitalia: ovipositor with moderately sclerotized papilla analis; apophyses moderate, long; ostium bursae a sclerotized cup with a central cone on intersegmental membrane between segments 7 and 8, or less distinct and in proximity to ductus bursae enlargement; ductus bursae thin, membranous, usually interrupted before bursa by enlargement for ductus seminalis juncture; corpus bursae ovate, with smaller accessory bursa anteriorly; signum on main bursal sac, a line of fused spicules or more diffuse spicule patch; ductus seminalis arising from enlarged ductus bursa section; bulla seminalis small. Larva: prolegs vestigial. Pupa: not protruded.

Remarks.- Neomachlotica is related to Machlotica, Abrenthia, and Trapeziophora, and is largely a Neotropical group with only one species entering North America in

southern Florida. Two species, currently undescribed, are
known from northeastern Mexico and may eventually be collected in subtropical southeastern Texas. Wing venation and
the unusual characters of the male and female genitalia
form the main differences from related genera. Head
morphology also shows features that isolate the genus,
notably the position of the anterior tentorial pits and unusual
maxillary palpi.

The only biological information known is that the single
Florida species feeds on a member of the Urticaceae as a
terminal bud and stem borer (Needham, 1955).

The following Neotropical species have been transferred
to **Neomachlotica** (Heppner, 1981b):

    Neomachlotica actinota (Walsingham, 1914) (Glyphip-
      teryx [sic])
    Neomachlotica atractias (Meyrick, 1909, (Machlotica)
    Neomachlotica nebras (Meyrick, 1909) (Machlotica).

Together with the Florida species described below, there
are 4 described species now assignable to this genus. I have
seen 3 additional undescribed species from Mexico and the
West Indies that belong in **Neomachlotica**.

Neomachlotica spiraea Heppner, 1981
(Figs. 3, 60–61, 135–137, 205–207)

Machlotica n. sp.- Needham, 1955:351.
Neomachlotica spiraea Heppner, 1981b:481, 1982a:45,
    1982b: 261.

Male (Fig. 60).- 3.2-4.0 mm. forewing length. **Head:**
dark fuscous with purple iridescence; frons with buff along
clypeal and lateral edges; labial palpus with basal segment
buff, 2nd segment dark fuscous with 2 buff transverse lines
ventrally and buff dorsally, and apical segment fuscous dorsally and dark fuscous ventrally with buff lateral borders;
maxillary palpus with very elongate middle segment; antenna
with fuscous dorsal scales. **Thorax:** fuscous; patagia fuscous
with purple iridescence; venter white; legs fuscous with
white at joints. **Forewing:** dark fuscous with approximately
14 narrow dotted longitudinal striae of greenish-yellow from
base to middle of wing, with distal end of striae-field
convex; dorsal margin near base with yellow scale line;
middle of wing with distally convex fuscous fascia, bordered
distally by a short silver fascia from the costal margin and
greenish-yellow scale striae towards tornus; middle silver

fascia from costa followed by yellow longitudinal striae and
then by two more short silver fascia toward apex; apex and
part of termen with silver border; fringe fuscous; ventrally
gray fuscous.   **Hindwing:** gray fuscous basally merging to
fuscous near termen; fringe fuscous; ventrally buff-gray
merging to fuscous near apex, with 3 short silver   fascia
from costal margin near apex similar to apical fascia of
forewing.   **Abdomen:** fuscous with silvery scales on posterior
of each segment; venter mostly white. **Genitalia** (Fig. 135):
tuba analis long, wide; tegumen rounded; vinculum rounded,
convex, without saccus; valva elongate, oblong, with rounded
apex, setaceous; valva with corematal setae on anterior side;
base of valva extended as unique narrow transtilla, overlapping with same from each valva; anellus short, tubular;
aedeagus (Fig. 136) short, nearly subequal to valval length,
narrow, without phallobase; aedeagus tip (Fig. 137) bulbous
with wide ring of stout recurved hooks surrounded by spicule
hood; cornutus a short tubule; ductus ejaculatorius from base
of aedeagus, with campanulate hood (2 preparations examined).

Female (Fig. 61).- 3.2-3.5 mm. forewing length. Similar
to male. **Genitalia** (Fig. 207): ovipositor short; papilla analis
sclerotized with sharp, incurving tip; apophyses long, thin,
with posterior pair slightly longer than anterior pair; ostium
bursae (Fig. 205) a shallow cup with a central cone having a
very small ostial opening, all in a larger oval depression
bordered laterally by two semi-circular ridges; ductus bursae
long and as thin as ostial opening, to sclerotized bulbous
area near bursa where ductus seminalis emerges; corpus bursae ovate with an accessory bursa half its size anteriorly
attached by a short duct; signum (Fig. 206) a linear fused
spicule line on bursa ventrum, half as long as longer bursal
diameter (4 preparations examined).

Larva.- pale, with black head capsule and prothoracic
tergal plate; prolegs vestigial.

Pupa.- Unknown.

Type.- Holotype male: Fisheating Creek, 2 mi. [=3.2
km] SE. Palmdale, Glades Co., Florida, 6 May 1975, on
flowers Pluchea purpurascens, J. B. Heppner (USNM). Paratypes (3 males, 5 females): **Florida.-** Dade Co.: Florida City,
21 Feb 1954 (2 females), M. O. Glenn (USNM); 25 Feb 1954
(1 male), M. O. Glenn (USNM).   Glades Co.: same data as
holotype, (1 female (JBH).   Highlands Co.: Archbold Biological Sta., (10 mi. [=16 km.] S. Lake Placid), 12 Jan 1965
(1 male), S. W. Frost (CPK); 23 Jan 1979, (1 male), H. V.
Weems, Jr., and S. Halkin (FSCA); 3 Mar 1952, reared ex

Boehmeria cylindrica (emerged 27 Mar 1952, 2 females), J. G. Needham (USNM). (Paratype to BMNH).

Additional specimens (3 females).- **Florida.**- Monroe Co.: Garden Key, Dry Tortugas, 8 May 1961 (3 males), R. E. Woodruff (CPK and FSCA).

Distribution (Fig. 3).- Known only from central to southern Florida.

Flight period.- January to March; May.

Host.- **Boehmeria cylindrica** (Linnaeus) Swartz (Urticaceae).

Biology.- The species has been reared by Needham (1955) but no larval or pupal specimens are available to give more detailed descriptions of these stages. The larva is a terminal bud borer, entering the terminal portion of the stem as well. Needham (1955) noted that a gall is formed where the larva feeds extensively in the stem. The terminal bud of the host, together with young leaves, are tied with silk. Pupation occurs near the bud or away from the host plant in a protected area. The cocoon is a fluted filigreed structure of amber silk which, from the description by Needham (1955), indicated a close resemblance to the cocoon of Neotropical **Ussara** species.

More recently the adults have been collected while feeding on flowers of **Pluchea purpurascens** (Swartz) de Candolle (Compositae) in an open cypress swamp along Fisheating Creek, Glades County, Florida.

Remarks.- The specimens available for study show no marked variations in wing pattern or coloration. As noted under the generic discussion, there are several species in Mexico, Central and South America, that superficially are very similar to **N. spiraea** in wing markings. The genitalia have the generic characteristics but are distinct specifically. The specimens of **N. spiraea** from the Dry Tortugas are too poor to make paratypes. Until the species is reared again and larvae and pupae are preserved, no morphological details of the immature stages will be known.

Drymoana Heppner, new genus
(Figs. 24-25, 32, 38, 44, 49, 54)

Type-species: **Drymoana blanchardi** Heppner, new species.

This genus is distinguished by the large scale tuft on the labial palpi and by the approximate veins Rs and M1 of the hindwings at the end of the cell in conjunction with the presence of M3.

Adults small (6.0–7.5 mm. forewing length). **Head** (Figs. 24, 32): frons smooth-scaled; vertex with scales loosely appressed; labial palpus recurved, with large ventral scale tuft on segment 2 (as long as segment 2) and segment twice length basal segment; apical segment with small ventral scale tuft and sharply acute, 2/3 length of 2nd segment; maxillary palpus (Fig. 25) 4-segmented; haustellum (Fig. 38) developed; pilifers small; ocellus (Fig. 49) moderate; eye moderately large; antenna (Fig. 44) long, with fine setae ventrally in males. **Thorax:** smooth-scaled. **Forewing** (Fig. 54): very elongate, with long pterostigma; costal and dorsal margins convex, meeting at acute apex, without distinct tornus; chorda developed; no vein in cell; Sc to costal margin before 1/2; R1–R4 to costal margin; R5 to apex; M1–M3 equidistant at base; M3 close to CuA1 at base; CuA2 divergent from CuA1; CuP at margin; A1+2 with short basal stalk; A3 vestigial. **Hindwing:** very elongate; costal margin somewhat convex; apex rounded; tornus broadly rounded, indistinct; dorsal margin straight to abrupt corner of anal margin; Sc+R1 nearly to apex; Rs to slightly before apex, approximate with M1 at base; vestigial vein in cell from Rs and M1 approximation point; M2 and M3 parallel in convex curve; M3 approximate to CuA1 at base; CuA1 and CuA2 parallel; CuP at margin; A1+2 with long basal fork, straight; A3 short; A4 at anal margin. **Abdomen:** posterior segment modified in male as genitalia hood with ventral split; no coremata. **Male genitalia:** tuba analis well-developed; tegumen stout; vinculum broad, with narrow saccus; valva simple, weak, with numerous fine setae on mesal side; valva with strongly sclerotized basal transtilla process forming base of anellus; anellus a long setaceous tube with aedeagus attached at distal end; aedeagus elongate with phallobase and apical setal band; cornutus a short tube. **Female genitalia:** ovipositor elongate, normal; papilla analis weak; apophyses very long and thin; ostium a membranous funnel on intersegmental membrane between segments 7 and 8 of ventrum; ductus bursae sclerotized to bursa; corpus bursae ovate; signum a spicule line; ductus seminalis from near ostium; bulla seminalis small. **Larva:** unknown. **Pupa:** unknown.

Remarks.- The genus is monobasic at present and includes one unusual species known only from the southeastern United States. The nearest relatives of the genus as based on genital characters appear to be **Glyphipterix** species of the **loricatella** species-group from Europe (based on **Glyphipterix loricatella** Treitschke). Although some **Glyphipterix** have developed a small scale tuft on the labial palpi, it is much more largely developed in **Drymoana** than in any **Glyphipterix**. The more fundamental characters of the genitalia and wing venation indicate a divergence from **Glyphipterix** to warrant the formation of this new genus.   No biological information is known thus far for the genus.

Drymoana is derived from the Greek for "forest dweller."

Drymoana blanchardi Heppner, new species
(Figs. 3, 62–63, 138–139, 208–209)

This species is distinguished by the fuscous and amber forewings with several silvery spots or fascia and the two apical fascia of the hindwings.

Male (Fig. 62).- 6.0–6.8 mm. forewing length.   **Head:** fuscous with a purple sheen; labial palpus with basal segment and base of 2nd segment fuscous, then 4 white bands alternating with dark fuscous to apex of apical segment; scale tuft fuscous with some white intermixed; antenna dark fuscous with purple sheen, white spot on scape and dorsally on 1st segment; antennal segments with silvery scales on mesad dorsum of each segment.   **Thorax:** fuscous; patagia fuscous; venter silvery white; legs dark fuscous with white on apex of each tarsal segment and at base of spurs of hind legs.   **Forewing:** fuscous at base and along costa; brown-buff on rest of wing; 8 silver iridescent spots in 4 pairs in dorsal half from base to 1/2, with distal pair much larger; 3 costal silvery iridescent bars with minute silver iridescent spot at 2/3 and white on costal margin above the 2 silver bars nearest apex; subterminal silver iridescent line; apex with silver iridescent bar; tornus with several minute silver iridescent spots; fringe fuscous, ending beyond tornus; venter fuscous mixed with white, costal white spots repeated and faint white line in place of dorsal subterminal silver line.   **Hindwing:** lustrous fuscous; 2 faint white fascia on distal 1/3; faint apical white spot; fringe fuscous; venter fuscous with white marks repeated but bolder.   **Abdomen:** fuscous with golden-yellow scales on posterior of each segment; anal tuft fuscous with golden-yellow posterior fringe.   **Genitalia** (Fig. 138): tuba analis very long and prominent; saccus short; valva with

acute saccular point and oblique distal end to convex dorsal
margin; tubular anellus long; aedeagus (Fig. 139) as long as
sacculus, with wide band of setae at tip and short tubular
cornutus (2 preparations examined).

Female (Fig. 63).- 7.0-7.5 mm. forewing length. As
described for the male but with more fuscous on forewing
and silver iridescent markings reduced in number but enlarged in size, with central cluster of 3-4 minute spots in
basal 1/4, large oblique fascia at near 1/2 extending across
wing, and large tornal bar across half of wing, with costal
and apical markings as in male but wider. Genitalia (Fig.
209): ovipositor weak; apophyses subequal, with anterior pair
somewhat stouter than posterior pair; ostium funnel deep;
ductus bursae as long as bursa length, sclerotized to bursa
and ending with a dorsal protrusion and carina; corpus bursae
oblong-ovate with sinuate, long spicule line for signum (Fig.
208) (2 preparations examined).

Types.- Holotype male: Deutschburg, Jackson Co., Texas,
7 Oct 1974, A. & M. E. Blanchard (USNM). Paratypes (6
males, 8 females): Florida.- Alachua Co.: Gainesville, Oct.
1983 (1 male) J. B. Heppner (FSCA). Escambia Co.: Paradise
Beach, 12 Oct 1949 (1 female, S. S. Nicolay (LACM); Pen-
sacola, 25 Sep 1962 (1 female), 5 Oct 1961 (1 male), S. O.
Hills (CPK). Gulf Co.: St. Joseph St. Pk., Cape San Blas, 2
Oct 1983 (1 male), J. B. Heppner (JBH). Highlands Co.:
Archbold Biol. Sta., (10 mi.[16 km.] S. Lake Placid), 8 Oct
1964 (2 females), P. H. Arnaud, Jr. (CAS). St. Johns Co.:
Hastings, 29 Sep (1 female), W. D. Kearfott (USNM). Volusia
Co.: Cassadaga, 18 Oct 1966 (1 female), S. V. Fuller (FSCA).
North Carolina.- Moore Co.: Southern Pines, 8-15 Sep (1
male) (USNM). South Carolina.- Charleston Co.: McClell-
anville, 17 Aug 1975 (1 female), 3 Oct 1973 (1 male), R. B
Dominick (RBD). Texas.- Jackson Co.: Deutschburg, 7 Oct
1974 (1 male), A. & M. E. Blanchard (AB). Montgomery Co.:
Camp Strake, 9 Sep 1975 (1 female), A. & M. E. Blanchard
(AB). (Paratype to USNM).

Distribution (Fig. 3).- North Carolina to Florida to
eastern Texas.

Flight period.- August to October.

Hosts.- Unknown.

Biology.- The biology is unknown but the host appears to
be a member of the southern pine belt understory. Only an
early autumn, partially late summer generation is known.

Remarks.- There is little variation in wing markings in
the few specimens available for study. The species is
relatively dimorphic in that the females have much larger

silvery iridescent markings on the forewings than do the males.

I have found specimens mixed in with laspeyresiine tortricids, mainly due to a superficial resemblance with **Cydia toreuta** (Grote). Careful examination will, of course, distinguish **D. blanchardi** as not being a tortricid.

The species is named in honor of Andre Blanchard, who first brought this species to my attention.

Glyphipterix Hübner
(Figs. 26-27, 33-34, 39-40, 45-46, 50-51, 55, 56, 132)

Glyphipterix Hübner, [1825]:421 (type-species: **Glyphipterix linneella** sensu Hübner, [1825] [not Clerck, 1759] (=**Tinea bergstraesserella** Fabricius, 1781), by subsequent designation, ICZN [2115], 1984:252); Fletcher, 1929:99; Möbius, 1936:133; Fletcher, 1946:128; Ford, 1949:126; 1954:97; Inoue, 1954:50; Turner, 1955:163: Empson, 1956:12; Inoue, et al., 1959:273; Moriuti & Saito, 1964:60; Bradley, 1965:105; Common, 1966: 42; Bradley & Pelham-Clinton, 1967:126; Clarke, 1969:56; Common, 1970b:811; Krogerus, et al., 1971:17; Bradley, 1972:11; Goater, 1974: 51; Watson & Whalley, 1975:211; Diakonoff, 1976:83; Diakonoff & Arita, 1976:179; Heath, 1976:102; Karsholt & Nielson, 1976:25; Alford, 1977:171; Diakonoff, 1977b:172; Diakonoff & Heppner, 1977:81; Diakonoff, 1978:45; 1979:303: Leraut, 1980:81; Heppner, 1981a:310; 1981b:479; 1982a: 46; 1982b:252; 1983a:25; 1984:54; Kyrki, 1984:78.
Heribeia Stephens, 1829:207 (type-species: **Tinea forsterella** Fabricius, 1781, by subsequent designation, Westwood, 1840a:112); Rennie, 1832:202 (nom. nud.); Stephens, 1834:262; Wood, 1837:194; Westwood, 1840a:112; 1845:210; 1854:194; Fletcher, 1929:107.
Aechmia Treitschke, 1833:69 (type-species: **Tinea fyeslella** Fabricius, 1798 (misspelling for **Tinea fueslella** Fabricius, 1781) [=**Phalaena thrasonella** Scopoli, 1763], by subsequent designation, Westwood, 1840a:112); Sodoffsky, 1837:22; Duponchel, 1838:434; Zeller, 1839a:181; 1839b:325; Westwood, 1840a:112; Duponchel, 1841:136; Fischer von Röslerstamm, 1841:233; Duponchel, [1845]:

360; Bruand d'Uzelle, 1850:48; Herrich–Schäffer, 1854:93; Stainton, 1854b:176; Speyer, [1856]:262; Desmarest, [1857]:282; Mann, 1857:181; Fre, 1858:138; Stainton, 1859:365; Möschler, 1866:143; Stainton, 1869:64; Staudinger, 1870:270; Fletcher, 1929:6; Kuznetsov & Stekolnikov, 1977:28.

Aecimia Boisduval, 1836:138, missp.

Glyphipteryx (Curtis, 1827:152, emend.); Zeller, 1839a:181, emend.; 1839b:325; Westwood, 1840b: 408; Duponchel, [1845]:361; 1845:243; Braund d'Uzelle, 1850:48; Stainton, 1854a:104; 1854b: 173; Newman, [1856]:281; Desmarest, [1857]: 282; Stainton, 1859a:364; 1859b:153; Walker, 1864:838; Stainton, 1869:202; Heinemann, 1870: 393; Stainton, 1879:228; Wocke, 1871:309; Morris, 1872:135; Stainton, 1872:xi; Barrett, 1874:69; Bang-Haas, 1875:35; Chambers, 1875b:292; Felder & Fogenhofer, 1875:10; Chambers, 1876:218; Millière, [1876]:347; Wocke, [1876]:393; Chambers, 1878b:148; Staudinger, 1879:353; Chambers, 1880a:64; Frey, 1880:380; Hartmann, 1880:93; Meyrick, 1880:226; Bang-Haas, 1881:204; Chambers, 1881:291; Walsingham, 1881:320; Donckier de Donceel, 1882:138; Fettig, 1882:166; Snellen, 1882:749; Curo, 1883:68; Jourdheuille, 1883:190; Jordan, 1886:154; Glaser, 1887:246; Moore, [1887]:524; Meyrick, [1888]:86; Riley, 1891:104; Meyrick, 1895:703; Morris & Kirby, 1896:135; Anonymous, 1897:255; Walsingham, 1897a:48; 1897b:119; Fereday, [1898]:367; Reutti, 1898: 180; Turner, 1898:204; Caradja, 1899:208; Seebold, [1899]:318; Dyar, 1900:84; King, 1901:254; Malloch, 1901:186; Rebel, 1901:130; Schütze, 1902:25; Dyar, [1903]:492; Kearfott, 1903:108; Hutton, 1904:122; Porritt, 1904:152; Lower, 1905:112; Crombrugghe de Piquerdaele, 1906:2; Bankes, 1907:204; Lameere, 1907:792; Meyrick, 1907:115; Nickerl, 1908:2; Walsingham, [1908]: 989; Meyrick, 1909:429; Spuler, 1910:298; Meyrick, 1911a:75; 1911b:291; 1912:56; 1913a:68; 1913b:23; Turner, 1913:211; Busck, 1914:61; Meyrick, 1914a:112; 1914c:28; Walsingham, 1914:300; Busck, 1915:87; Meyrick, 1915a:221; 1915b:225; Martini, 1916:113; Meyrick, 1916:418; Rebel, 1916:79; Barnes & McDunnough, 1917:181; Janse, 1917:203; Meyrick, 1918:195; Philpott, 1918:129;

Meyrick, [1920]: 1005; 1920a:296; 1920b:331; 1921a:113; 1922:489; Bauer, 1923:167; Forbes, 1923:355; Braun, 1925:204; Chopra, 1925:359; Handlirsch, 1925:878; Tillyard, 1926:422; Turner, 1926:144; Fletcher, 1927:25; Hering, 1927:432; Lycklama, 1927:13; Philpott, 1927:346; Turner, 1927:157; Zerny, 1927:471; Hudson, 1928:305; Meyrick, 1928:708; Vorbrodt, 1928:116; Waters, 1928:252; Amsel, 1930:116; Issiki, 1930:426; Matsumura, 1931:1079; Meyrick, 1931:184; Schütze, 1931:26; Vorbrodt, 1931:122; Esaki, et al., 1932:1485; Hering, 1932:176; Eckstein, 1933:108; Sterneck & Zimmerman, 1933:79; Meyrick, 1935: 86; Pierce & Metcalfe, 1935:41; Randou, 1935: 226; Amsel, 1936:353; Toll, 1936:404; Le Marchand, 1937a:192; Amsel, 1838:113; Hudson, 1939:456; McDunnough, 1939:84; Turner, 1939: 114; Comstock, 1940:633; Rebel, 1940:37; Fletcher & Clutterbuck, [1941]:104; Jäckh, 1942:194; Turner, 1942:93; Heslop, 1945:27; Kloet & Hinks, 1945:132; Le Marchand, 1945:126; Fletcher, 1946: 128; Procter, 1946:318; Thompson, 1946:262; Diakonoff, 1948:206; Lhome, 1948:499; Amsel, 1949:88; Herbulot, 1949:117; Amsel, 1950:26; Bourgogne, 1951:386; Hartig & Amsel, 1951:92; Börner, 1953:397; Gozmány, 1954:277; Wörz, 1954:84; Hartig, 1956:126; Hering, 1957:341; Amsel, 1959:8; Moruiti, 1960:16; Klimesch, 1961: 724; Kodama, 1961:41; Jones & Jones, 1964:91; Bleszynski, et al., 1965:413; Kimball, 1965:287; Legrand, 1965:49; Davies, 1973:206; Hannemann & Urbahn, 1974:320; Heppner, 1974:292.

Glyphiteryx Fischer von Röslerstamm, 1741:233, missp.

Anacampsoides Bruand d'Uzelle, 1850:32 (type-species: **Heribeia simpliciella** Stephens, 1834, by monotypy), nom. nud.

Glypipteryx Stainton, 1854a:104, missp.

Glyphopteryx Herrich-Schäffer, 1854:92, missp.

Glyphiptoryx Mann & Rogenhofer, 1878:500, missp.

Glyphptieryx Turati, 1879:203, missp.

Glyphipterys Christoph, 1882:38, missp.

Glyphyteryx Hampson, 1918:387, missp.

Glyphteryx Watt, 1920:439, missp.

Adults small (3.4–9.2 mm. forewing length). Head (Figs. 26, 33–34): frons and vertex smooth-scaled; labial palpus recurved, dorso-ventrally flattened, generally smooth-scaled, sometimes with scale tufts, apical segment subequal to 2nd segment or to 1.5 times longer, pointed; 2nd and basal segments nearly subequal; maxillary palpus (Fig. 27) short, sometimes long enough to partially overlap haustellum, 4-segmented; pilifer moderate; haustellum (Figs. 39–40) developed; ocellus (Figs. 50–51) moderate; eye usually moderate, rarely large; antenna (Figs. 45–46) moderate, fine setae ventrally in males. Thorax: smooth-scaled; patagia short. Forewing (Figs, 55–56): elongate–oblong; costal margin somewhat convex or nearly straight, apex acute, often falcate but rounded; termen oblique, rounded convexly to indistinct tornus; dorsal margin convex; anal angle rounded; pterostigma long; Sc to costal margin at 1/2; R1–R4 to costal margin, R5 to termen; rarely R4 and R5 stalked; chorda developed, sometimes vestigial; spur of vein in cell; M1–M3 usually evenly spaced, free; M3 emergent from base of CuA1 or more distant, parallel to CuA1, upcurved or straight; CuA2 close to CuA1, nearly straight to tornus; CuP at tornus, extended as fold; A1+2 with long basal stalk; A3 at anal margin. Hindwing: elongate, lanceolate with distal end often broader than basal 1/3 or acute at apex with indistinct termen; usually 3/4 as long as forewing; costal margin somewhat sinuate, to abruptly rounded apex; termen oblique to convex tornus; dorsal margin straight to rounded anal angle; Sc+R1 to before apex; Rs to apex, sometimes vestigial near base; M1+M2 close together, with central vein through cell; M3 from base of CuA1 or more distant, convergent with CuA1 at termen; CuA2 close to CuA1, nearly parallel to CuA1; CuP at tornus, extended as fold; A1+2 with long basal fork, curved toward margin; A3 distant from anal margin; A4 vestigial. Abdomen: posterior segment modified in male as genitalia hood, with venter split; lateral coremata on posterior segment or absent. Male genitalia: tuba analis well-developed, often laterally somewhat sclerotized, usually long; tegumen stout, usually entire, sometimes split dorsally; vinculum broad, convex; saccus an elongate narrow extension, sometimes reduced, rarely very stout or absent, sometimes with extended rods to valval bases through vinculum; valva simple, setaceous, rarely with corematal setae on anterior side, sometimes variously modified and strongly sclerotized; anellus (Fig. 132) tubular, with connection to basal transtilla process of valva; aedeagus elongate, thin, often with apical spicule collar; phallobase short or reduced; cornutus a short tubule; vesica often with spicules; ductus ejaculatorius with a hood near aedeagus. **Female genitalia:** ovipositor short or long; papilla analis usually simple, setaceous, sometimes sclerotized with sharp edges; apophyses usually long and thin, sometimes stout; ostium bursae usually a membranous funnel or cup on intersegmental membrane between sternites 7 and 8, sometimes variously sclerotized or modified; ductus bursae usually thin and long, membranous or sclerotized, sometimes short; ductus seminalis from ductus bursae or bursa near ductus bursae junction; bulla seminalis small; corpus bursae usually ovate, often tear-drop shaped, sometimes with an accessory bursa on anterior end; signum usually absent, rarely present as spicule area. **Larva** (Figs. 269-279: **G. semiflavana**): head with fronto-clypeus reaching to epicranial notch; 2 adfrontal setae (1 in **G. semiflavana?**); 6 stemmata in semi-circle; prothorax with L-group bisetose; thoracic legs developed; mesothorax with one SV seta; abdominal segments with D1 closer together than D2 but subequally distant on segment 6; L1 directly dorsad of spiracle on abdominal segments; segments 9-10 sometimes with tergal plates, sometimes with long setae on segment 10; lateral projections on segment 10; spiracles on produced cylinders or integumental projections, largest on prothorax and abdominal segment 8; prolegs vestigial; crochets absent. **Pupa** (Figs. 289-293: **G. semiflavana**): elongate with horn-like projections on head or reduced, with lateral projecting spiracles on prothorax; appendages to wing tips; two small setae on each tergite; no distinct cremaster but ventral and posterior hook-tipped setae.

Remarks.- **Glyphipterix** is distinguished from closely related genera by the complete venation, thus not lacking M3 in the hindwings, and in the usually smooth-scaled, long labial palpi. There are no consistently distinctive features of the genitalia at the generic level but males usually have simple valvae and a narrow saccus, while females usually have a simple cup-shaped ostium and lack a signum; these characters, together with wing venation, enable the genus to be defined.

The genus includes 276 species throughout the world, including the species newly described herein, and is fairly evenly distributed throughout the faunal regions, although with more species described from areas that have been more thoroughly collected. The most typical members of the genus have a white crescent mark on the dorsal margin of the forewing, although many species lack this mark and have

various other markings. Most all species, however, have at
least some metallic iridescent markings on the forewings.
The hindwings are invariably unmarked, usually fuscous.

In North America north of Mexico 27 species are recorded in the genus. The North American fauna has some
Neotropical elements, notably in the **circumscriptella**
species-group, and a high number of unusual endemic, often
archaic species (**brauni, quadragintapunctata,** and **bifasciata**
species-groups), but most of the species have considerable
relationships to the Palearctic fauna. Superficially many
species greatly resemble species from the Palearctic, but
there is only one known Holarctic species, **Glyphipterix
haworthana** (Stephens).

The genus is rather large and although future research
may indicate groupings at the subgeneric level, many of the
exotic species are too poorly known at the present time to
discern any clear divisions in the genus. **Glyphipterix** has
not been informally subdivided in the past except by Meyrick
(1914c) and some other authors like Spuler (1910), whose
groupings include species showing little discernible resemblance among each other except on maculation to some
extent. Thus, the use of species-groups defined by a wider
range of morphological and phenotypic characters is an innovative feature of this revision. Although the remainder of
the species beyond the Nearctic are not at present arranged
into species-groups, a few are mentioned in relation to their
affinities to North American species or where members of
extralimital faunas belong to the groups defined for North
American species.

There has been some question of the type-species of
Glyphipterix Hübner due to confusion with a type-species
selection of Westwood (1840a), which now refers to the
genus **Glyphipteryx** sensu Curtis and now used for **Chrysoclista** in Agonoxenidae; the details of all this are noted in
Diakonoff and Heppner (1977) and ICZN (1984). The major
problem, however, is the recent use of the Curtis generic
name instead of **Chrysoclista,** which causes great confusion
with the emendation of **Glyphipterix** Hübner to **Glyphipteryx.**

Biologies are known for only a few species of the world
fauna and these were summarized previously in the section
on glyphipterigid biology. As noted previously, most known
Glyphipterix species have larvae that are either seed, stem
or bud borers, or leaf miners, thus far mostly in monocotyledonous plant families. Larval characters have in the
past been accurately presented only by authors dealing with
particular species (Chopra, 1925; Empson, 1956; Moriuti,

1960; Kodama, 1961). Larval descriptions supposedly for the family (Common, 1970b; Peterson, 1965) refer exclusively to Choreutidae.

The key to species based on superficial adult wing characters is difficult to use for the **montisella** species-group (couplets 19-26) because the species are so similar; most of these species will have to be dissected for comparison of genital characters for correct identification. In the key to species based on male genital characters the last couplet is also difficult to use without comparative material.

Key to Glyphipterix Species Based on Maculation

1.   Forewing with distinct field of longitudinal striae on apical quarter or over midwing 2
         Forewing with various markings but no striae field as above 8

2(1).   Striae field very large, covering most of midwing (Figs. 64-65) ....
         .... **quadragintapunctata** (p. 74)
         Striae field small, restricted to apical quarter..........3

3(2).   Forewing dorsal margin with large vertical white fascia near base extended across wing or nearly to costal margin 4
         Forewing dorsal margin with small white fascia (never reaching beyond cubitus) or without fascia or mark 5

4(3).   Dorsal margin white fascia extended to costal margin, narrow; a second white fascia across wing near midwing (Figs. 74-75) ....
         .... **bifasciata** (p. 85)
         Dorsal margin white fascia not reaching costal margin, broad on dorsal margin; midwing without white fascia extended across wing but with 2 silver bars from costal margin and 1 silver bar from costal margin (Figs. 79-81) ....
         .... **circumscriptella** (p. 71)

5(3).   Forewing without basal white mark on dorsal margin near base (Fig. 83) ....**brauni** (p. 69)
         Forewing with small basal white mark on dorsal margin base 6

6(5).   Costal margin of forewing with small oblique white mark on basal 1/4 (Fig. 78) .....**hypenantia** (p. 88)
Costal margin lacking white mark on basal 1/4....... 7

7(6).   Forewing with 3rd white mark from apex on costal margin extended with silver surrounded by buff; generally smaller than G. **yosemitella** (Figs. 76–77) ....**unifasciata** (p. 91)
Forewing with 3rd white mark from apex on costal margin extended with silver but not surrounded by buff; generally larger than typical G. **unifasciata** (Fig. 82) ....**yosemitella** (p. 90)

8(1).   Forewing with dorsal margin with 2 large white marks, one basal and one near tornus or midwing 9
Forewing with one large white mark on dorsal margin, near tornus or midwing 17

9(8).   Forewing costal margin with 6 or 7 white marks ...10
Forewing costal margin with 5 white marks (rarely with minute 6th mark on margin; if present then not surrounded by fuscous border) (**californiae** species-group) 12

10(9).   Forewing costal margin with 7 white marks (**powelli** species-group 11
Forewing costal margin with 6 white marks (Figs. 114–115) .... **ruidosensis** (p. 131)

11(10).   Midwing dorsal margin white mark nearly quadrate (Figs. 70-73 .... **urticae** (p. 81)
Midwing dorsal margin white mark triangular or rectangular (Figs. 66–69) ....**powelli** (p. 77)

12(9).   Basal dorsal margin white mark with recurved extension along cubitus (Figs. 92-93); head buff ....
.... **californiae** (p. 102)
Basal dorsal margin white mark without recurved mesal extension; head fuscous 13

13(12).   Basal dorsal margin white mark triangular, usually rounded on distal mesad edge of mark (Figs. 100-101) .... **arizonensis** (p. 111)

Basal dorsal margin white mark quadrate-trapezoid-
al, broad or narrow, but not triangular 14

14(13). Basal dorsal margin white mark narrow, sinuate, like
a very slanted Z (Figs. 96–97) 15
Basal dorsal margin white mark broad, quadrate-
trapezoidal. 16

15(14). Basal dorsal margin white mark with reduced distal
end (Figs. 98–99) (this species should be checked
in genitalia keys) ....sierranevadae (p. 109)
Basal dorsal margin white mark with extended distal
end (Figs. 96–97) (This species should be checked
in genitalia keys) ....feniseca (p. 105)

16(14). Costal margin with basal crescent mark not reaching
convergence of next mark with dorsal midwing
mark (Figs. 90–91) ....roenastes (p. 112)
Costal margin with basal crescent mark reaching
convergence of next mark with dorsal margin
midwing mark (Figs. 94–95) ......juncivora (p. 106)

17(8). Dorsal margin with large white crescent at midwing
(haworthana species-group) 18
Dorsal margin with large white mark distad of
midwing and relatively straight, not a curved and
pointed crescent (montisella species-group) 19

18(17). Costal margin with 4th white mark from apex extended to tornus as a chevron with silver (Figs. 82–
83) ....haworthana (p. 94)
Costal margin with 4th white mark with silver
mesad but not extended to tornus (Figs. 86–89) ..
....sistes (p. 98)

19(17). Anal margin with distinct white line to base of
forewing 20
Anal margin without white line to base of forewing..
....26

20(19). Anal margin white line relatively broad, as broad as
apical white marks 21
Anal margin white line relatively thin, not as broad
as apical white marks 22

21(20).   Costal margin of forewing with 6 white marks (Figs.
          102-103) (Arizona) .... chiricahuae (p. 115)
          Costal margin with 5 white marks (if 6 marks then
          one is minute and without fuscous border) (Fig.
          106) (eastern) .... chambersi (p. 122)

22(20).   Dorsal margin white mark distad of midwing trapez-
          oidal, not very pointed mesad (Fig. 104) ....
          .... saurodonta (p. 119)
          Dorsal margin white mark distad of midwing with
          pointed mesal end, somewhat triangular 23

23(22).   Costal margin with 6th white mark from apex prom-
          inent and distinct to convergence of next white
          mark with dorsal mark (Figs. 110-111) 24
          Costal margin with 6th white mark from apex not
          wider than other costal marks and weakly ex-
          tended to convergence of next mark with dorsal
          mark or not extended 25

24(23).   Forewing basal 1/3 with large yellow-buff oval spot
          (Fig. 107) ....flavimaculata (p. 126)
          Forewing basal 1/3 fuscous without basal spot (at
          most a small pale area but not a large distinct
          spot) (Figs. 110-111) .... montisella (p. 123)

25(24).   Forewing apical 1/2 with buff scaling over fuscous
          ground color; basal costal mark 6th from apex
          curved weakly to convergence of next mark with
          dorsal mark (Figs. 108-109) (Arizona) ....
          .... hodgesi (p. 117)
          Forewing apical 1/2 with dark fuscous; basal costal
          mark 6th from apex straight, not extended all
          the way to convergence of next mark with dorsal
          mark (Fig. 105) (Tenn./N.C.) ..... cherokee (p.120)

26(19).   Costal margin white mark 6th from apex curved and
          extended to convergence of next mark from dor-
          sal margin (Figs. 112-113) ....santaritae (p.129)
          Costal margin white mark 6th from apex straight
          and not extended to convergence of next mark
          with mark from dorsal margin (Fig. 116) ....
          .... melanoscirta (p. 127)

Key to **Glyphipterix** Species Based on Male Genitalia[3]

1.   Valvae with corematal setae on distad anterior side
        (seen as a non-setaceous oval area in genital
        preparations after corematal setal fall away) ...2
        Valvae without corematal setae on distad anterior
        side 6

2(1).   Seccus prominent, stout, elongate (Fig. 140).............
        .... quadragintapunctata (p. 74)
        Saccus reduced, appearing absent...........................3

3(2).   Valvae oblong with distal ends evenly rounded.........4
        Valvae oblong with distal ends truncated, at least
        ventral apex quadrate if not whole termen
        truncate 5

4(3).   Valvae with distal 1/2 little broader than middle
        (Fig. 142) (California) .... powelli (p. 77)
        Valvae with distal 1/2 noticeably broader than mid-
        dle (Fig. 146) (Great Basin and Rockies) ....
        .... urticae (p. 81)

5(3).   Valvae with distal end only somewhat truncated, but
        ventral apex quadrate (Fig. 144) (Washington) ...
        .... powelli jucunda (p. 77)
        Valvae with distal end truncated (Fig. 148) (Alberta
        and Manitoba) ....urticae sylviborealis (p. 83)

6(1).   Valvae with apex extended as a thin elongate point
        (bifasciata species-group) 7
        Valvae various but without a thin elongate point on
        apex 10

7(6).   Aedeagus with long apical spicule collar (Fig. 161)...
        .... yosemitella (p. 90)
        Aedeagus with short apical spicule collar 8

---

[3] The males of **Glyphipterix chambersi**, G. **cherokee**, and
G. **saurodonta** are unknown.   The males of the **bifasciata**
species-group are virtually identical and the characters used
in the key (couplets 6-9) can only be used adequately in
conjunction with relative sizes of specimens for comparison;
the illustrations of these species give some aid for relative
size (Figs. 150-155, 160-161).

8(7).   Aedeagus 3/4 valval length (Fig. 152).........................
           .... hypenantia (p. 88)
           Aedeagus 2/3 valval length 9

9(8).   Aedeagus moderately narrow (Fig. 151) ....
           .... bifasciata (p. 85)
           Aedeagus very thin (Fig. 155).........unifasciata (p. 91)

10(6).   Valvae elongate, rather narrow 11
           Valvae very oblong with broad distal end................13

11(10).   Saccus reduced, appearing absent (Fig. 187)..............
           ....sistes (p. 98)
           Saccus distinctly present, narrow 12

12(11).   Valvae with middle distinctly broader than base or
           apex (Fig. 158) ....brauni (p. 69)
           Valvae with most of length of uniform width, with
           small apical point (Fig. 185) ...haworthana (p. 94)

13(10).   Valvae with triangular setaceous distal end, without
           projected point (Fig. 156) circumscriptella (p. 71)
           Valvae with round setaceous distal end or very mod-
           ified and sclerotized 14

14(13).   Valvae short, very sclerotized with large extended
           point from saccular apex and dense spines on
           costal apex (Fig. 189) ....ruidosensis (p. 131)
           Valvae not modified as above, weakly sclerotized,
           setaceous, with rounded costal apex and termen .
          15

15(14).   Valvae without projected points, somewhat truncated
           on central termen (Fig. 173) ...roenastes (p. 112)
           Valvae with small or large projected point from sac-
           cular apex 16

16(15).   Saccular point with valvae very short (shorter than
           saccus) 17
           Saccular point of valvae long (longer than saccus or
           equal) 22

17(16).   Saccular point distinctly pointed ventrad (Fig. 171)...
           .... arizonensis (p. 111)
           Saccular point distinctly pointed distad................18

18(17). Saccular margin relatively straight.........................19
         Saccular margin noticeably convex.........................20

19(18). Saccus long (2/3 tuba analis length) (Fig. 162)..........
         ....flavimaculata (p. 126)
         Saccus short (1/3 tuba analis length) (Fig. 169).........
         ....sierranevadae (p. 109)

20(18). Valvae somewhat truncated distally with sharply
         convex costal edge (Fig. 165) ...juncivora (p. 106)
         Valvae rounded on distal end.........................21

21(20). Aedeagus short (subequal to valval length) (Fig. 168).
         ....feniseca (p. 105)
         Aedeagus long (subequal to distance from tuba analis
         tip to saccus tip) (Fig. 176) ....
         ....californiae (p. 102)

22(16). Saccular point directed straight ventrad (Fig. 179)....
         ....hodgesi (p. 117)
         Saccular point directed in line with saccular margin
         (distad) or oblique 23

23(22). Saccular point oblique (Fig. 177) ....
         ....chiricahuae (p. 115)
         Saccular point in line with saccular margin...........24

24(23). Invagination of valval termen next to saccular point
         relatively small (1/2 saccus length) (Fig. 181) ....
         ....montisella (p. 123)
         Invagination of valval termen next to saccular point
         large (nearly subequal to saccus length) 25

25(24). Genitalia small relative to abdominal size...............
         ....melanoscirta (p. 127)
         Genitalia large relative to abdominal size ....
         ....santaritae (p. 129)

Key to Glyphipterix Species Based on Female Genitalia[4]

1. Ovipositor with papilla analis strongly sclerotized....2
Ovipositor with papilla analis not strongly sclerotized 4

2(1). Papilla analis elongated, rounded, without piercing edges (Fig. 211) ....quadragintapunctata (p. 74)
Papilla analis pointed, with piercing edges (Figs. 251, 253) 3

3(2). Ostium bursae a funnel with an anterior sclerotized cup (Fig. 256); bursa copulatrix without a signum
.... haworthana (p. 94)
Ostium bursae a funnel without any sclerotization (Fig. 253); bursa with a spicule–patch signum (Fig. 252) ....sistes (p. 98)

4(1). Ostium bursae a membranous funnel without any anterior sclerotization 5
Ostium bursae a sclerotized cup, often elongated, or at least with the major portion of the cup sclerotized 9

5(4). Ostium bursae with an elongated funnel (more than twice width of ovipositor tip in length (Fig. 226)
......................................circumscriptella (p. 71)
Ostium bursae with funnel very short (less than width of ovipositor tip in length) (powelli species-group) 6

6(5). Anterior apophyses shorter than posterior pair (Fig. 215) .... powelli jucunda (p. 77)
Anterior and posterior apophyses subequal...............7

7(6). Posterior apophyses subequal to anterior pair anterior to basal fork (Fig. 219) ....
................................urticae sylviborealis (p. 83)
Posterior apophyses longer than anterior pair anterior to basal fork 8

_____

[4] The females of Glyphipterix flavimaculata, G. melanoscirta, and G. yosemitella are unknown.

8(7).   Posterior apophyses with truncated ends (Fig. 213)...
.... **powelli** (p. 77)
Posterior apophyses with small spatulate ends (Fig.
217) .... **urticae** (p. 81)

9(4).   Ostium with anterior sclerotized cup; accessory bursa absent 10
Ostium fully sclerotized; accessory bursa present ..17

10(9). Ostium elongate with small anterior sclerotized cup.
.... 11
Ostium short (dimensions subequal), with small
anterior sclerotized cup 13

11(10). Posterior apophyses with spatulate ends (Fig. 223)....
.... **hypenantia** (p. 88)
Posterior apophyses with truncated ends without any
widening 12

12(11). Ostium anterior cup with roughened pattern in sclerotization (Fig. 220) .... **bifasciata** (p. 85)
Ostium anterior cup without any pattern in the
sclerotization (Fig. 224) ....**unifasciata** (p. 91)

13(10). Posterior apophyses with expanded truncations on
ends (Fig. 239) ....**californiae** (p.102)
Posterior apophyses without expanded truncations on
ends 14

14(13). Anterior apophyses spatulate (Fig. 230) ....
.... **feniseca** (p.105)
Anterior apophyses truncated................................. 15

15(14). Sclerotized cup of ostium small, narrow, cone-shaped
(Fig. 234) ....**arizonensis** (p. 111)
Sclerotized anterior cup of ostium larger than in **arizonensis**, rounded near ductus bursae or a broad
cone-shape 16

16(15). Anterior cup of ostium a broad cone (Fig. 228).........
....**juncivora** (p. 106)
Anterior cup of ostium a rounded cone (Fig. 232).....
....**sierranevadae** (p. 109)

17(9).   Bursa elongate with small accessory bursa as small
         knob (Fig. 237) ....roenastes (p. 112)
         Bursa ovate with round accessory bursa.....................
        18

18(17). Ostium a long cup; ovipositor with ventral sclerotiz-
         ed projections on 8th sternite (Fig. 258) ....
         ....ruidosensis (p. 131)
         Ostium short; ovipositor without modification as
         above 19

19(18). Ostium with length-width dimensions subequal.......20
         Ostium with length longer than width dimension....21

20(19). Ostium very small with broad rounded anterior end
         (Fig. 254) ....cherokee (p. 120)
         Ostium larger with somewhat pointed anterior end
         (Fig. 256) ....chambersi (p. 122)

21(19). Sternite 7 with distinct posterior edge (Fig. 248)......
         ....saurodonta (p. 119)
         Sternite without distinct posterior edge.................22

22(21). Ostium nearly twice longer than wide (Fig. 240).......
         ....chiricahuae (p. 115)
         Ostium with length not more than 1.5 times width...
        23

23(22). Ostium with length only somewhat more than width,
         broadly rounded anterior end (Fig. 246) ....
         ....santaritae (p. 129)
         Ostium with length distinctly longer than width.....24

24(23). Ostium a broad cone-shaped cup, appearing
         triangular (Fig. 242) ....hodgesi (p. 117)
         Ostium a narrow cone-shaped cup with more rounded
         anterior end (Fig. 244) ....montisella (p. 123)

The brauni species-group

Forewing with falcate indentation below rounded apex,
with elongate-oblong. Hindwing oblong with distinct termen.
Labial palpus short, ventrally roughened with scales, not very
dorso-ventrally flattened; scales of frons long, extended over
haustellum base. Eye large. Abdomen without coremata.

Valva elongate, unmodified; tegumen split dorsally; saccus narrow, moderately long. Female unknown.

The species-group currently contains only one unusual species from Wyoming.

The short labial palpus and the large eye are unusual features in **Glyphipterix**. The species appears to be archaic member of the genus, possibly having closest relatives in south Asia. **Glyphipterix enclitica** Meyrick, from Sri Lanka, may belong to this species-group, since it appears to be very similar both in wing maculation and genitalia to the North American species. In the Nearctic fauna the species-group is isolated and relatively distant even from the possible closer relatives in the **circumscriptella** species-group.

Glyphipterix brauni Heppner, new species
(Figs. 4, 83, 158, 159)

The forewings resemble several other species in the genus but the short labial palpi and the genitalia are distinctive.

Male (Fig. 83).- 6.2 mm. forewing length. **Head:** fuscous, mixed with buff on frons, scales extended beyond clypeus; labial palpus short, dorsally buff fuscous, venter white basally, fuscous on 2nd and apical segments, apical segment shorter than normal in genus; antenna fuscous dorsally. **Thorax:** fuscous; patagia fuscous; venter fuscous; legs fuscous, white bands at tarsal joints. **Forewing:** fuscous ground color, overlaid with dense brown scales from basal 1/3 of wing; large apical 1/3 buff ovate area near tornus, with 3 short black longitudinal striae in center; costal margin with white fascia oblique to cubitus near base; white fascia at midwing, somewhat broken centrally by small silver spot; basally oblique fascia basad of buff area of tornus, ending at a silver spot; apex with two short white bars from costal margin, ultimate one almost merging with white at tornal fringe; tornus with 5 black spots alternating with 5 silver spots, with the basad silver spot elongated as bar to 2nd adjacent black spot on cubitus; fringe fuscous and white on apex, white from falcate indentation to tornus; venter fuscous with apical 2 white fascia repeated, silver spots repeated as faint marks; fringe repeated from dorsum. **Hindwing:** dorsum, fringe and venter fuscous. **Abdomen:** fuscous, some silver scales on posterior of each segment; venter similar but with white. **Genitalia** (Fig. 158): tuba analis very long; tegumen split, thin; vinculum narrow, convex; saccus short, narrow; valva very elongate, basally narrow, setaceous; short tubular anellus, setaceous apically; aedeagus (Fig. 159) thin, short, 1/2 length valva; short spicule band apically; cornutus a short tubule; vesica with spicules; phallobase indistinct, rounded; small hood on ductus ejaculatorius near aedeagus (1 preparation examined).

Female.- Unknown.

Type.- Holotype male: Grand Teton National Park, [Teton Co.], Wyoming, 6 Jul 1959, A. F. Braun (ANSP).

Distribution (Fig. 4).- Known only from Wyoming.

Flight period.- July.

Hosts.- Unknown.

Biology.- Unknown.

Remarks.- This species is based on a unique specimen from Wyoming. It is an interesting species inasmuch as its nearest relatives appear not to be Nearctic but rather Oriental. Glyphipterix enclitica Meyrick from Sri Lanka may be more closely related to it than any other known species. The labial palpi of G. enclitica are also somewhat short and the genitalia are quite similar in both species. The wing maculation of G. brauni is similar to Glyphipterix loricatella (Treitschke) from Europe and species of the bifasciata species-group from North America, as well as the circumscriptella species-group, yet in all these other species the genitalia are more specialized and the labial palpi are both long and tufted ventrally.

The species is named in honor of Dr. Annette F. Braun, the collector of this species and many other glyphipterigids.

The circumscriptella species-group

Forewing with somewhat sharp apex, followed by falcate indentation, wing elongate-oblong. Hindwing oblong with distinct termen. Labial palpus relatively smooth-scaled ventrally, apical segment somewhat short. Eye moderate. Abdomen without coremata. Valva elongate, distally greatly expanded; tegumen entire; saccus narrow. Ovipositor long; papilla analis simple, setaceous; apophyses long and thin; ostium bursae a membranous funnel; ductus bursae short, membranous; corpus bursae tear-drop shaped; signum absent.

In North America this species-group contains only Glyphipterix circumscriptella Chambers. There are numerous Neotropical species and probably some in the Palearctic that belong in this species-group. The forewing markings of these species all are very similar among themselves and the North

American species. The group is perhaps closest to the **quad-ragintapunctata** species group of North America.

Glyphipterix circumscriptella Chambers

Glyphipteryx [sic] circumscriptella Chambers, 1881: 291; Riley, 1891:104; Dyar, [1903]:493; Kearfott, 1903:108; Morse, 1910:553; Meyrick, 1913b:46; 1914c:33; Barnes & McDunnough, 1917:182; Forbes, 1923:355; Leonard, 1928:554; McDunnough, 1939:84.

Glyphipteryx [sic] circumscripta Dyar, 1900:84, missp.

Glyphipterix circumscriptella.- Heppner, 1982a:48; 1983a:26.

This species is easily distinguished from similar species in North America by the genitalia, while the forewing dorsal margin white mark is unique in the North American fauna.

Glyphipterix circumscriptella circumscriptella Chambers, new status
(Figs. 4, 80–81, 156–157, 226–227)

Male (Fig. 80).- 4.2–4.5 mm. forewing length. **Head:** lustrous fuscous, with thin buff eye margin; labial palpus dorsally white, venter smooth-scaled, white with central fuscous line from base to apex; antenna fuscous dorsally. **Thorax:** lustrous fuscous; patagia fuscous; venter white and fuscous; legs fuscous with white at joints; spurs white. **Forewing:** ground color dark fuscous, with brown scales overlaid on some areas of distal 2/3 of wing; dorsal margin with large triangular white mark at 1/3 wing length from base, reaching to radius; at midwing silver bar to cell from dorsal margin and opposite two corresponding silver bars from costal margin, with each bar having a white spot on the costal margin; apical field a fuscous triangle with 7–8 longitudinal scale striae of buff; short silver crescent oblique from falcate indentation; apex with a white wedge-shaped mark, silver at acute tip of mark; tornus a black wedge with 3 silver spots interrupted or partially surrounded by several small dots of buff; basad silver spot of tornus with adjacent mesal silver spot; fringe dark fuscous, gray distally, interrupted by white at falcate indentation; venter fuscous with dorsal silver marks indistinct buff and white marks repeated. **Hindwing:** fuscous with white border along termen; fringe fuscous; venter similar. **Abdomen:** fuscous with white scales on posterior each segment; venter mostly white. **Genitalia** (Fig. 156): tuba analis long; tegumen entire, very narrow dorsally; vinculum rectangular, convex; saccus long, narrow; valva setaceous on enlarged distal ends, basally narrow, dorsal margin sharply convex after basal area; short tubular anellus, finally setaceous; aedeagus (Fig. 157) long, thin, subequal to valva; phallobase small, rounded; cornutus a tube 1/2 length aedeagus with apical invagination with numerous small spicules; vesica with fine spicules; ductus ejaculatorius with hood near aedeagus (2 preparations examined).

    Female (Fig. 81).- 5.0-5.8 mm. forewing length. Similar to male. **Genitalia** (Fig. 227): ovipositor long; papilla analis simple, setaceous; apophyses long and thin, posterior pair longer than anterior pair; ostium bursae (Fig. 226) an indistinct membranous funnel merging into very short ductus bursae; ductus bursae sclerotized at bursa entrance; ductus seminalis from ductus bursae entrance to bursa; corpus bursae tear-drop shaped, large; signum absent but with numerous spicules over most of bursa (2 preparations examined).

    Type.- Holotype male: "1028"; "1322". [Massachusetts: Amherst, Hampshire Co.] (USNM).

    Additional specimens (10 males, 7 females).- **Connecticut.-** Fairfield Co.: South Shore, 11 Jul 1930 (1 female), A. B. Klots (AMNH). **Illinois.-** Putnam Co.: 21 Jun 1949 (1 male), M. O. Glenn (USNM). **New Jersey.-** Essex Co.: Essex Co. Park, 1 Jul (2 females), W. D. Kearfott (USNM); 7 Jul (1 male), W. D. Kearfott (USNM). **Texas.-** Bosque Co.: Laguna Park, 29 Apr 84 (2 males), E. C. Knudson (FSCA). **Canada.-** **Ontario.-** Leamington, 24 Jun 1931 (1 male), 27 Jun 1931 (1 male), 2 Jul 1931 (1 female), 3 Jul (1 male, 1 female), W. J. Brown (CNC); Toronto, Jun 1913 (1 female), Jul 1916 (1 male), Jul 1922 (1 male, 1 female), Aug 1912 (1 male), Parish (BMNH).

    Distribution (Fig. 4).- Records are from Massachusetts, Connecticut, New Jersey, southern Ontario, Illinois and Texas.

    Flight period.- Late June to July; August.

    Hosts.- Unknown.

    Biology.- Unknown.

    Remarks.- Although widely distributed in the northeastern United States and Canada, **G. circumscriptella**

is rare in collections, probably being very localized.   A
presence for the species in Texas was only recently discovered. Variation is evident only in slight size differences
in the white mark of the dorsal margin of the forewing.

Glyphipterix circumscriptella apacheana Heppner, new
subspecies
(Figs. 4, 79)

An Arizona race which has the fuscous areas of the
forewings gray in appearance and certain apical scales white
instead of buff.

Male (Fig. 79).- 5.0 mm. forewing length. Similar to the
nominate subspecies with the following differences: forewing
with fuscous scales white tipped after basal area, giving a
gray appearance, with brown-tipped scales on apex; large
apical 1/3 fuscous area with white longitudinal striae; black
tornal border with yellow by middle silver spot, remaining
border spots around silver spots are white; large dorsal
margin white mark narrower than in nominate subspecies.
Genitalia: indistinguishable from nominate subpsecies (1
preparation examined).
Female.- Unknown
Types.- Holotype male: Stewart Camp, 1 mi. [=1.6 km.]
S. Portal, [Chiricahua Mts.], Cochise Co., Arizona, 26-29
Aug 1971, ex Malaise trap, J. T. Doyen (UCB). Paratype (1
male): Arizona.- same data as holotype, 23-25 August 1971
(1 male) (UCB). (Holotype to CAS).
Distribution (Fig. 4).- Known only from southeastern
Arizona.
Flight period.- August.
Hosts.- Unknown.
Biology.- Unknown.
Remarks.- This population has male genitalia virtually
identical with the eastern populations, thus, until the female
is discovered it is not known whether it represents a distinct
species. Many related species from South America are also
very similar, both phenotypically and in genital morphology
in males.
This subspecies is named after the Apache Indians of
Arizona.

The quadragintapunctata species-group

Forewing without falcate indentation below acute apex, wing relatively broad, elongate–oblong. Hindwing oblong with distinct termen. Labial palpus with venter smooth–scaled, apical segment very long. Eye moderate. Abdomen without coremata. Valva setaceous, with large corematal area on anterior side; tegumen entire; vinculum strongly sclerotized; saccus very stout but long and narrow. Ovipositor short; papilla analis strongly sclerotized laterally, setaceous; apophyses stout, thin, with large basal fork on anterior pair; ostium bursa a membranous cup; ductus bursae moderate, thin; ductus seminalis emergent from bursa near ductus bursa; corpus bursae oval; signum absent.

Only one species is currently recognized in this group. Although superficially similar to the European loricatella species–group, the North American group differs in wing venation, presence of coremata on the valvae, stout papilla anales, and smooth–scaled labial palpi.

Glyphipterix quadragintapunctata Dyar
(Figs. 5, 33, 39, 45, 50, 55, 64–65, 140–141, 210–211)

Glyphipteryx [sic] quadragintapunctata Dyar, 1900: 84; [1903]:493; Kearfott, 1903:108; Meyrick, 1913b:45; 1914c:32; Barnes & McDunnough, 1917: 182; Forbes, 1923:355; McDunnough, 1939:84.
Glyphipterix quadragintapunctata.- Heppner, 1982a: 52; 1983a:26.

The combination of yellow forewing markings on a partially black fuscous field and the distinctive genitalia make this species easily recognized. Although G. brauni is superficially similar to G. quadragintapunctata, the genitalia are very different.

Male (Fig. 64).- 5.4–5.6 mm. forewing length. Head (Figs. 33, 39, 50): fuscous vertex, merging to buff on frons and lateral edges; labial palpus dorsally buff, ventrally buff mixed with fuscous except fuscous ventral tip of apical segment; antenna (Fig. 45) fuscous dorsally. Thorax: fuscous; patagia buff with some fuscous mixed in; ventrally buff; legs with some fuscous on tarsal segments. Forewing (Fig. 55): fuscous on basal 1/5 with white spot on costal margin; bordering basal fuscous area is vertical fascia of silver iri-

descence from costal margin to near anal margin where end
of fascia is cream white, merging into yellow anal margin at
base; following basal fascia is fuscous area by dorsal margin
and a broad black wedge–shaped area to tornus with numerous irregular spots of yellow and 4 spots of silver iridescence
in trapezoidal arrangement near tornus; silver iridescent
fascia broken in middle at midwing from costal margin to
dorsal margin, with white end spot on costal margin; yellow
along costa to apex and termen from basal fascia; beyond
midwing are 4 white costal spots, continued as silver iridescent fascia except 3rd from apex; 4th white spot from
apex with silver fascia oblique to sharp bend near tornus;
2nd white spot from apex with short silver fascia and 1st
white spot from apex with silver fascia twice length white
spot; termen edged with narrow fuscous line; fuscous border
often on costa and interrupted by white spots; fringe silvery
fuscous; venter fuscous with white near terminal fascia from
apex. **Hindwing:** fuscous, fringe white with fuscous base,
becoming fuscous along anal margin; venter fuscous overlaid
with white scales. **Abdomen:** fuscous with silvery white
scales on posterior of each segment; venter mostly white
with some fuscous. **Genitalia** (Fig. 140): tuba analis elongate; tegumen narrow band basally, wide posteriorly on dorsum; vinculum very stout, somewhat truncated postero-
ventrad, with abrupt junction to narrow but stout saccus;
valva elongate, dorsally rounded to distal ventral blunt point,
with large corematal area on anterior side, setaceous overall; short tubular anellus; aedeagus (Fig. 141) thin, subequal
to valval length, with apical 1/3 with spicule collar; short
phallobase; cornutus a short tubule; ductus ejaculatorius with
hood near aedeagus (4 preparations examined).

Female (Fig. 65).- 5.8–6.8 mm. forewing length. Similar
to male. **Genitalia** (Fig. 211): ovipositor short; papilla analis
sclerotized, long, setaceous; apophyses thin, with posterior
pair somewhat longer than anterior pair and strong posterior
fork with bend on anterior pair; ostium bursae (Fig. 210) a
circular membranous cup merging into a narrow membranous
ductus bursae; ductus bursae as long as bursal diameter;
ductus seminalis emergent from bursa near ductus: corpus
bursae ovate; signum absent (3 preparations examined).

Type.- Holotype female: Onaga, [Pottawatomie Co.],
Kansas, [no date], Crevecoeur (USNM, type 4424).

Additional specimens (7 males, 7 females).- Illinois.-
Putnam Co.: 15 Jun 1956 (2 males, FMNH; 2 males, 1 female
USNM), M. O. Glenn; 16 Jun 1956 (1 female, FMNH; 1 fe-
male, USNM), M. O. Glenn; 29 Jun 1957 (1 female), M. O.

Glenn (FMNH); [no date] (1 female), [M. O. Glen], A. K.
Wyatt Coll. (FMNH). Kansas.- Pottawatomie Co.: Onaga, [no
date] (1 male, 1 female, MCZ; 1 male, USNM), [Creve-
coeur?]. Ohio.- Adams Co.: Beaver Pond, 11 Jun 1930 (1
male), A. F. Braun (ANSP); Mineral Springs, 27 Jun 1931 (1
female), A. F. Braun (ANSP).

Distribution (Fig. 5).- Collection records are from Ohio,
Illinois, and Kansas.

Flight period.- June.

Hosts.- Unknown.

Biology.- Unknown.

Remarks.- Some specimens show minor variations in the
extend of varous fascia or spots, but generally they are quite
uniform. The male genitalia of G. quadragintapunctata are
very unusual in the genus due to the very stout and thickened saccus, as well as the presence of valval coremata. It
is unfortunate that this species, together with the previous
two species and also species like G. loricatella in Europe,
are unknown biologically, since these are among the most
unusual species in the genus and are also inferred to be
among the least advanced in the genus.

The powelli species-group

Forewing with falcate indentation below apex, wing
elongate-oblong.   Hindwing oblong with distinct termen.
Labial palpus somewhat roughened ventrally; apical segment
not much longer than 2nd segment. Eye moderate. Abdomen without coremata. Valva setaceous, simple, corematal area on anterior side; tegumen entire; vinculum often
with tubular extension centrally from reduced or long saccus.
Ovipositor short; papilla analis simple, setaceous, slightly
sclerotized; apophyses short, thin, with long basal fork on
anterior pair; ostium bursae a membranous cup; ductus bursae membranous, thin; ductus seminalis from ductus bursae
or at bursa junction; corpus bursae tear-drop shaped, large,
or oval; signum absent.

This species-group contains two Nearctic species that
form a sibling species group in western North America. I
have chosen to treat them as two species rather than as
subspecies of one widely dispersed species primarily because
of genitalic differences that indicate reproductive isolation,
since the genitalia are quite uniform for the one species
(Glyphipterix urticae, new species) for which enough material
is available to provide a sample of several disjunct

populations within its known range.   Thus, the apparently
minor character differences observed in the genitalia appear
to refer to valid, though perhaps only recently evolved,
species.

The group is characterized by the valval coremata, the
lack of a long saccus in most species (undescribed related
species from Mexico have a long saccus), and female genitalic characters and wing maculation.   There may be some
relationship to the **bifasciata** species-group but nearest
relationships apparently are to the **quadragintapunctata**
species-group, since both of these groups have the extension
of the saccus into the vinculum sclerite and possess valval
coremata, which the **bifasciata** species-group does not
possess.

Glyphipterix powelli Heppner, new species

The forewing markings together with the short posterior
apophyses of the female distinguish this species from near
relatives.

Glyphipterix powelli powelli Heppner, new subspecies
(Figs. 6, 66-67, 142-143, 212-213)

Male (Fig. 66).- 3.9-6.2 mm. forewing length.   Head:
fuscous, becoming buff on frons; labial palpus cream white
dorsally, venter with basal segment white, 2nd segment
fuscous with buff distally, and apical segment fuscous;
antenna with fuscous dorsally.   Thorax: fuscous, some buff
posteriorly; patagia fuscous with some buff laterally; venter
white and fuscous; legs fuscous with white rings at joints.
Forewing: ground color dark fuscous near base and as borders
to most wing markings; yellow-brown scales scattered over
remaining areas not silver or white; dorsal margin with white
narrow fascia 1/4 wing length from base and another more
triangular fascia at midwing, oblique toward apex, reaching
CuP fold and with silver spot distally; costal base with white
spot; distad to basal buff spot, costal margin with 7 short
white bars of varying length to apex, most with silver dots
on mesal ends; white bar 4th from apex extended as silver
chevron, ending at tornus white spot; 2 silver spots near end
of cell; silver spot below basal white costal fascia near mid-
terminal white indentation of wing falcation; small white
spot at tornus before white of silver chevron; fringe black
fuscous, distally with white at falcate indentation; ventrally

lustrous fuscous; white spots repeated but small and more
sharply defined. **Hindwing:** lustrous fuscous, somewhat paler
toward base; fringe long, fuscous, with white distally along
termen; venter lustrous fuscous overlaid with white scales
and white near apex and on anal field. **Abdomen:** fuscous
with white scales on posterior of each segment; venter similar but mostly white. Genitalia (Fig. 142): tuba analis long;
tegumen extended to point as broad dorsal hood; vinculum
sharply rounded, convex; saccus reduced to short point; valva
elongate, somewhat oblong, distally rounded, setaceous, with
anterior corematal setae; short tubular anellus; aedeagus
slightly curved, short, about 3/4 valval length; aedeagus (Fig.
143) without apical spicule hood; cornutus a short tubule;
bulbous phallobase not distinctly demarcated; ductus ejaculatorius with hood near aedeagus (5 preparations examined).

Female (Fig. 67).- 5.8-7.5 mm. forewing length. Similar
to male. Genitalia (Fig. 213): ovipositor short; papilla analis
slightly sclerotized, elongate; apophyses long, fairly stout,
subequal; anterior apophyses curved, with large posterior
fork; ostium bursae (Fig. 212) a small membranous cup merging with funnel of ductus bursae; ductus bursae narrow,
short; ductus seminalis at ductus bursae entrance to bursa;
corpus bursae very oblong, gradually widening from ductus
bursae diameter to ovate anterior end; no signum (2 preparations examined).

Types.- Holotype male: Dune Lakes, San Luis Obispo Co.,
California, 24 Feb 1975, J. A. Powell (UCB). Paratypes (13
males, 3 females): California.- Martin Co.: Inverness, 15 Mar
1959 (1 male), 5 Apr 1959 (1 male, 1 female), D. Burdick
(UCB). San Luis Obispo Co.: Dune Lakes, 24 Feb 1975 (7
males, 1 female), J. A. Powell (UCB). Santa Cruz Co.:
Santa Cruz, 27 Mar 1961 (1 male), R. Brown (CAS); 5 Apr
1961 (2 males), R. Brown (CAS); 16 Apr 1933 (1 male), J. C.
Elmore (CDAS). Sonoma Co.: Cazadero, 14 Apr 1918 (1 female), J. C. Bradley (CU). (Holotype to CAS; paratypes to
BMNH, FSCA and USNM).

Additional specimens (1 male).- "Califor"[nia], [no date]
(1 male), Coll. H. Edwards (AMNH).

Distribution (Fig. 6).- Known only from the California
coastal range.

Flight period.- February to April.

Hosts.- **Urtica** sp.? (Urticaceae).

Biology.- The species has not been reared but adults
have been collected on **Urtica**, which may be the larval host.

Remarks.- Fresh specimens have the covering scales of
yellow-brown on the forewings over most of the wing, while

older specimens often have lost some of the scales, giving
them a darker appearance. The white dorsal markings vary
somewhat, especially the midwing dorsal margin fascia which
can be somewhat narrow or broad. The one additional specimen not designated as paratype is in very poor condition.

The slightly oblong valva and the interior smooth edge of
the tegumen are characters distinguishing **G. powelli** from
the other species of the complex in North America. In **G.
powelli jucunda**, new subspecies, the valvae are more pronounced oblong, larger, and the interior edge of the tegumen
is pointed, while the female has longer posterior apophyses
than the other species. In **G. urticae**, new species, the wing
markings of the dorsal margin are usually much more quadrate and the basal one usually has an extension going to the
costal buff spot, the valvae are more narrow basally, the
tegumen often has an interior point, and the aedeagus is
more curved. The female bursa also appears to be somewhat
larger in relative size than in **G. powelli**.   **Glyphipterix
urticae sylviborealis**, new subspecies, also has more quadrate
dorsal wing markings and the extension of the basal marking,
while the valvae are very truncate distally. The latter new
species (**G. urticae**) has genitalia generally smaller in the
male than do equally sized males of **G. powelli**.

Glyphipterix powelli is named in honor of Dr. Jerry A.
Powell, University of California, Berkeley, who collected
most of the specimens and has actively collected glyphipterigids for me during the course of this revision.

Glyphipterix powelli jucunda Heppner, new subspecies
(Figs. 6, 68-69, 144-145, 214-215)

This is a northwestern subspecies distinguished by its
large size, the broadly oblong valvae and the long posterior
apophyses. The lack of a clear forewing distal silver chevron is distinctive.

Male (Fig. 68).- 6.2 mm. forewing length. **Head:** fuscous,
with some buff near clypeus; labial palpus nearly white
dorsally, ventrally with basal segment white, 2nd segment
buff, and apical segment fuscous with 2 white and 2 buff
bands alternated towards base; antenna fuscous dorsally.
Thorax: fuscous with some buff posteriorly; patagia mostly
buff; venter white and fuscous; legs fuscous with white at
joints. **Forewing:** ground color dark fuscous on basal half as
borders of markings, covered in other areas with yellow-

brown scales; costal margin with basal white spot followed by white chevron fascia across wing but broken by fuscous at subcostal vein; dorsal margin with second large white, often nearly triangular oblique fascia at midpoint ending at CuP fold with a silver spot; costal margin with 7 white linear spots distad of basal chevron, all ending mesally with a silver spot or bar; large silver spot near end of cell, closely followed distally by two smaller silver spots, and then by 4 more irregular silver spots mesad of the penultimate costal white mark; silver spot at tornus; silver and white bar basad of tornus; fringe black-fuscous with fuscous distally and white at apical falcate indentation, buff at tornus, venter fuscous with white spots repeated and termen with white line from near apex to beyond tornus. **Hindwing**: lustrous fuscous; fringe fuscous with white distally on termen; venter fuscous overlaid with white scales over most of wing, especially the apex. **Abdomen**: fuscous with white scales on posterior of each segment; venter same but more white. **Genitalia** (Fig. 144): tuba analis long; tegumen broad and pointed dorsally, with interior projection; vinculum rounded, convex; saccus reduced to short point; valva elongate-oblong, simple, setaceous, with ventral distal corner nearly quadrate, anterior side with corematal setae; short tubular anellus; aedeagus (Fig. 145) short, 3/4 valval length, slightly curved, without apical spicule band; phallobase bulbous but indistinct; cornutus a short tubule; ductus ejaculatorius with hood near aedeagus (1 preparation examined).

Female (Fig. 69).- 6.8 mm. forewing length. Similar to male but with more buff on vertex. **Genitalia** (Fig. 215): ovipositor short; papilla analis slightly sclerotized, long; anterior apophyses stout with large posterior fork; posterior apophyses long and thin, twice length of anterior apophysis portion anterior to fork; ostium bursae (Fig. 214) a small membranous cup merged to a short, thin ductus bursae; ductus seminalis arising at ductus bursae junction to bursa; corpus bursae very elongate, oblong, gradually widening from ductus bursae width; no signum (1 preparation examined).

Types.- Holotype male: Pullman, [Whitman Co.], Washington, 26 May 1951, R. B. Spurrier (WSU). Paratypes (2 females): **Washington.**- Whitman Co.: Pullman, 27 May 1965 (1 female), R. D. Akre (WSU); 2 Jun 1965 (1 female), R. D. Akre (WSU). (Paratype to USNM).

Distribution (Fig. 6).- Known only from eastern Washington.

Flight period.- May to June.

Hosts.- Unknown.

Biology.- Unknown.

Remarks.- The subspecies has a distinct forewing broken chevron fascia near the base which is only an oblique fascia or complete to the costal margin in the related species. The silver spots are also irregular near the wing apex while in the other species there is a well-defined silver chevron.

The specific name is derived from Latin for "pleasant."

Glyphipterix urticae Heppner, new species

Glyphipteryx [sic] **montisella.**- Braun, 1925:205 (not Chambers) (misdetermination).

The quadrate dorsal marginal mark of the forewing midpoint together with the more narrow dorsal area of the tegumen, often with an interior point, will serve to distinguish this species.

Glyphipterix urticae urticae Heppner, new subspecies
(Figs. 6, 70-71, 146-147, 216-217)

Male (Fig. 70).- 4.2-6.0 mm. forewing length. **Head:** fuscous, with narrow border of buff along clypeal and lateral eye margins; labial palpus nearly white dorsally, venter of basal segment white, 2nd segment fuscous with buff band apically, and apical segment fuscous with buff near base; antenna fuscous dorsally. **Thorax:** fuscous, some buff posteriorly; patagia small, fuscous with some buff posteriorly; venter white and fuscous; legs fuscous with white at joints. **Forewing:** fuscous ground color predominant on basal half, bordering markings on apical half where most other areas are overlaid with yellow-buff scales; basal half overlaid with gray scales; dorsal margin with white bar fascia near base and white quadrate fascia, sometimes triangular or narrower at midpoint, extended to CuP fold; costal margin with buff spot beyond basal white spot, then 7 oblique white bars to apex, each with silver spot mesally in most cases; silver spot at end of cell; costal white spot 4th from apex extended to basad of tornus as silver chevron fascia, at tornus ending in a white spot; apex with a dark fuscous spot; white spot at tornus extended by short silver bar; fringe dark fuscous distally, white at falcate indentation, buff at tornus; venter fuscous with white marks sharply repeated; white on anal field. **Hindwing:** fuscous; fringe fuscous with black-fuscous base along termen; venter

fuscous with white at apex and on anal field. **Abdomen:**
fuscous with silvery scales on posterior of each segment;
venter fuscous with white scales on posterior of each segment and mostly white near thorax. Genitalia (Fig. 146):
tuba analis long; tegumen with narrow dorsal band, interior
edge often with pointed projection; vinculum sharply rounded,
convex; saccus reduced to short point; valva very elongate,
oblong, distal end rounded, setaceous, anterior side with
corematal setae; anellus short, tubular; aedeagus (Fig. 147)
short, 3/4 valval length, curved with no apical spicule band;
cornutus as short tubule; phallobase bulbous, indistinct;
ductus ejaculatorius with hood near aedeagus (8 preparations
examined).

    Female (Fig. 71).- 5.0-6.2 mm. forewing length. Similar
to male but with basal white fascia of dorsal margin extended to costal buff spot by very thin line of white.
Genitalia (Fig. 217): ovipositor short; papila analis slightly
sclerotized, long, setaceous; apophyses average, nearly
subequal, anterior pair with large posterior fork; ostium
bursae (Fig. 216) a small membranous cup merging as a funnel to short ductus bursae; ductus seminalis from ductus
bursae entrance to bursa; corpus bursae elongate-oblong,
gradually widening from ductus bursae width, anteriorly large
and ovate; signum absent (5 preparations examined).

    Types.- Holotype male: South Platte River, 1.5 mi. [=2.4
km.] SW. Lake George, Park Co., Colorado, 1 Jul 1976, 8200'
[=2500 m.], on **Urtica**, J. B. Heppner (USNM).   Paratypes
(39 males, 13 females): **Colorado.**- Park Co.: same data as
holotype, (21 males, 6 females) (JBH).   San Juan Co.: Silverton, 8-15 Jul (8 males, 3 females, USNM; 1 male, 1
female, ANSP), 16-23 Jul (1 male, USNM), [9200' (=2800
m.)].   **New Mexico.**- Lincoln Co.: S. Fork Bonito Cr. Cpgd.,
Sacramento Mts., 6 Jul 1977 (5 males, 1 female), 7600'
[=2320m.], on **Urtica**, J. B. Heppner (USNM).   Otero Co.:
Ruidoso Cyn., 30 Jun 1939 (1 female), A. F. Braun (ANSP)
Utah.- Rich Co.: Swan Cr., 29 Jun 1924 (2 males), A. F.
Braun (ANSP).   Utah Co.: Timpooneke Cpgd., Mt. Timpanogos, 29 Jul 1967 (1 male), 7400' [=2250 m] J. A. Powell
(UCB). (Paratypes to BMNH, CNC, FSCA, and USNM).

    Additional specimens (1 male).- **California.**- Mono Co.: 4
mi. E. Monitor Pass, 15 Jul 1966 (1 male), J. A. Powell
(UCB).

    Distribution (Fig. 6).- Collection records are from
Colorado, New Mexico, Utah, and California.

    Flight period.- Late June to late July.

    Hosts.- **Urtica** sp.? (Urticaceae).

Biology.- The species has not been reared but adults have only been collected close to **Urtica**, which appears to be the larval host. Populations appear to be localized, since many areas with extensive areas of **Urtica** were devoid of the moths but this may be due to varying times of emergence from one locality to another.

Remarks.- The basad white fascia of the dorsal margin is similar to the chevron fascia of G. **powelli jucunda** but the extension to the costa appears to be only present in females of G. **urticae**, often indistinct. **Glyphipterix urticae** differs markedly from the preceding species in having no yellow-brown overlaying scales on the basal half of the forewing.

The specimen from California is from an area having greatest faunal and floral relationships to the Great Basin, but due to the disjunct locality, the specimen is not added to the paratype series.

The specific name is derived from the name of the presumed host plant, **Urtica**.

Glyphipterix urticae sylviborealis Heppner, new subspecies
(Figs. 6, 72-73, 148-149, 218-219)

Male (Fig. 72).- 5.2-5.5 mm. forewing length. **Head**: fuscous with narrow border of buff laterally and along clypeus; labial palpus dorsally white, ventrally with basal segment white, 2nd segment with 2 alternating bands of fuscous and buff, and apical segment fuscous with some white near base; antenna dorsally fuscous. **Thorax**: fuscous with some buff scales; patagia fuscous with buff posteriorly; venter fuscous and white; legs fuscous and white at joints. **Forewing**: ground color dark fuscous, predominantly in basal half and always bordering markings; apical half, sometimes also basal half, with overlaid scales of yellow-brown; basal white spot on costa; dorsal margin with white bar near base extended as chevron to buff costal spot and another white quadrate bar at midwing, extending to CuP fold; distad of basal marks, costal margin with 7 white bars usually with mesal ends as silver spots; silver spot or spots at end of cell; silver chevron from costal white bar 4th from apex extended to white spot basad of tornus; silver spot by fringe area of white; fringe dark fuscous, distally fuscous, then white, interrupted by white at falcate indentation; venter fuscous with white spots repeated. **Hindwing**: lustrous fuscous; fringe fuscous, white distally along termen; venter fuscous overlaid with white scales; all white at apex and on anal field. **Genitalia (Fig. 148):** tuba analis long; tegumen wide dorsally, interior edge truncate; vinculum rounded, convex; saccus reduced as short point; valva elongate, distally truncated, setaceous, with corematal setae on anterior side; short tubular anellus; aedeagus (Fig. 149) short, 3/4 valval length, without spicule collar on apex, curved; indistinct bulbous phallobase; cornutus a short tubule; ductus ejaculatorius with hood near aedeagus. (2 preparations examined).

Female (Fig. 73).- 5.5-7.0 mm. forewing length. Similar to male; sometimes more yellow-brown overlaid scales on forewing. **Genitalia (Fig. 219):** ovipositor short; papilla analis slightly sclerotized, long, setaceous; apophyses thin, subequal, anterior apophyses with long posterior fork; ostium bursae (Fig. 218) a small round cup, membranous, merging into short funnel; short membranous ductus bursae; ductus seminalis from ductus bursae junction with bursa; corpus bursae very elongate, gradually widening from ductus bursae width to ovate anterior end; signum absent (2 preparations examined).

Types.- Holotype male: Waterton Lakes [Provincial Park], Alberta, Canada, 28 Jun 1923, J. H. McDunnough (CNC). Paratypes (3 males, 5 females): **Alberta.-** same locality as holotype, 28 Jun 1923 (1 male, 1 female), 30 June 1923 (1 female), 6 Jul 1923 (1 female), 8 Jul 1923 (1 female), J. H. McDunnough (CNC). **Manitoba.-** Aweme, 30 June 1922 (1 female), N. Criddle (CNC). **Saskatchewan.-** Cypress Hills, nr. Maple Creek, 3 Jun 1926 (2 males), C. H. Young (CNC). (Paratypes to USNM).

Distribution (Fig. 6).- Southern Manitoba to southern Alberta, Canada.

Flight period.- June to early July.

Hosts.- Unknown.

Biology.- Unknown.

Remarks.- Most of the available specimens lack yellow-brown scaling on the basal half of the forewing, although in good condition, and appear as dark as G. **urticae urticae,** thus, darker than G. **powelli.** The two specimens from Saskatchewan have yellow-brown scaling on the basal half of the forewing. The genitalia of the dissected specimens are uniform in the specific distinctions of the truncated valvae and subequal apophyses of the female.

The specific name is derived from Latin for "woods of the north."

The bifasciata species-group

Forewing elongate-oblong, with falcate indentation. Hindwing somewhat oblong, termen reduced. Labial palpus with relatively long scale tufts from base of 2nd segment to near sharp apex of ultimate segment in decreasing length, apical segment short. Eye moderate. Abdomen without coremata. Valva elongate with narrow pointed tip; tegumen split dorsally; saccus narrow and long, with Y-shaped connecting rods to valval bases. Ovipositor long; posterior apophyses long; papilla analis simple, setaceous; ostium bursae a deep membranous cup, sometimes basally sclerotized; ductus bursae thin and long, usually at least partially sclerotized; corpus bursae oval, small, without accessory bursa, signum absent.

This species-group is composed of 4 species which form a very compact group from western North America, predominately in the Cascade and northern Coast Ranges of the Pacific Coast. The genitalia of the species are extremely similar in the males, less so in the females. Although somewhat related to both the **powelli** and the **haworthana** species-groups, in addition to superficially appearing more related to G. **brauni**, there are no known species that are very closely related to this species-group.

Glyphipterix bifasciata Walsingham
(Figs. 7, 74-75, 150-151, 220-221)

Glyphipteryx [sic] bifasciata Walsingham, 1881:321; Riley, 1891:104; Dyar, 1900:84; [1903]:492; Kearfott, 1903:108; Meyrick, 1913b:45; 1914c:32; Barnes & McDunnough, 1917:182; McDunnough, 1939:84.
Glyphipterix bifasciata.- Heppner, 1982a:48; 1983a: 26.

The large size and the basal white fascia generally reaching the costal margin of the forewing will serve to distinguish this species.

Male (Fig. 74).- 6.0-7.6 mm. forewing length. Head: fuscous, with buff line along clypeus and lateral eye margins; labial palpus white and fuscous dorsally, venter with basal segment white; scale tufts long, from base of 2nd segment tapering to near apex, 4 alternating tufts of fuscous and white; antenna fuscous dorsally. **Thorax:** fuscous; patagia fuscous; venter white with some fuscous; legs fuscous with white at joints. **Forewing:** ground color fuscous; oblique white fascia from basal 1/4 of dorsal margin to costal margin or nearly there; vertical white fascia at midwing somewhat broken in cell by silver spot; apical 1/3 of wing with large field of white scales overlaying ground color, with silver, black bordered spot near tornus; silver spot at end of cell; costal margin with 3 white bars from 2/3 wing length to apex, all with silver spots mesally; falcate indentation with silver spot on wing; tornus with 5 silver spots encircled by solid black; fringe dark fuscous, distally white except gray along apex and white at falcate indentation; venter fuscous, white and silver dorsal markings faintly repeated with white scales. **Hindwing:** lustrous fuscous; fringe fuscous; venter fuscous overlaid with white scales, white line along termen. **Abdomen:** fuscous with white on posterior of each segment; venter similar but more white. **Genitalia** (Fig. 150): tuba analis very long; tegumen split dorsally; vinculum rounded, convex; saccus long and thin, often with bulbous apex; valva elongate, narrow, with long setae on mesal side, apex a long extended point; short tubular anellus; aedeagus (Fig. 151) long and narrow, 2/3 valval length, with spicule collar on tip; phallobase narrower than aedeagus; cornutus a short tubule; vesica with spicules; ductus ejaculatorius with large hood near aedeagus (10 preparations examined).

    Female (Fig. 75).- 4.9-7.9 mm. forewing length. Similar to male, sometimes forewings darker fuscous. **Genitalia** (Fig. 221): ovipositor long; papilla analis simple, setaceous; apophyses long and thin, posterior pair longer than anterior pair; ostium bursae (Fig. 220) a wide membranous funnel with a basal sclerotized cup; ductus bursae long, thin, sclerotized toward bursa; ductus seminalis from ductus bursae and bursa junction; corpus bursae oval, small; signum absent (10 preparations examined).

    Types.- Lectotype male, by present designation; Mt. Shasta, Siskiyou Co., California, 2 Aug-1 Sep 1871, Walsingham 92018 (BMNH). (Labelled additionally: Walsingham, Collection, 1910-427; TYPE male desc., fig[ure]d [Durrant label]; LECTOTYPE male, **Glyphipteryx bifasciata** Wlsm., By Heppner '76; B.M. male, Genitalia Slide, No. 20231). Paralectotypes (3 males, 1 female): same data as lectotype, (1 male, 1 female), Wlsm. 92019 & 92017 (BMNH); (1 male) (MCZ, type 14989); "26" [Bear Valley, Colusa Co., Cali-

fornia, 27 Jun 1871], (1 male), Walsingham, Stainton Coll.
401447 (BMNH).
   Additional specimens (87 males, 20 females).- **California.**- Mendocino Co.: N. fork Big R., 7 Jun 1871 (2 females),
Walsingham 92014-15 (BMNH). San Bernardino Co.: Fredalba,
25 Jul 1912 (1 male), 13 Aug 1912 (1 male), 19 Aug 1912 (1
male), G. R. Pilate (ANSP). Santa Cruz Co.: Lockhart
Gulch, 4 mi. [=6.4 km.] E. Mt. Herman, 25 Jul 1961 (1
female), D. J. Burdick (UCB). Shasta Co.: Big Bend, 3 Jul
1965 (1 female), 2800'[=855 m.], R. M. Brown (CAS). Siskiyou Co.: Fowlers Camp, 5 mi.[=8 km.] E. McCloud, 1 Jul
1963 (3 males), C. D. MacNeill & V. B. Whitehead (CAS), 22
Jul 1962 (1 female), D. C. Rentz and C. D. MacNeill (CAS);
7 Jul 1957 (3 males, 1 female), 13 Jul 1962 (11 males), 14
Jul 1962 (8 males), J. A. Powell (UCB); Shasta Retreat, 8-15
Jun (1 male, USNM), 16-23 Jun (5 males, USNM; 1 male,
LACM), 24-30 June (5 males, USNM; 1 male, LACM), 1-7 Jul
(2 males, USNM), [Barnes & McDunnough?]; Mt. Shasta City,
4 Jul 1958 (1 male), J. A. Powell (UCB); Winchester, 31 Jul-
1 Aug 1871 (1 male), Walsingham 92016 (BMNH). "Cal." (1
female), Walsingham (USNM). **Idaho.**- Moscow Mts., 20 Jul
1930 (1 female), 22 Jun 1930 (1 male), J. F. G. Clarke
(USNM). **Montana.**- Missoula Co.: Camela Hump Rgr. Sta.,
17 Jul 1926 (1 female), E. H. Nast (CAS). **Nevada.**- Washoe
Co.: Verdi, 20-30 Jun (1 male), A. H. Vachell (MCZ). **Oregon.**- Baker Co.: Baker, Spring Cr., 25 Jul 1963 (1 female),
J. H. Baker (USNM); 2 mi [=3.2 km.] NW. Sumpter, 18 Jun
1970 (2 males), J. F. G. Clarke (USNM). Grant Co.: Ritter,
17 Jul 1962 (1 female), 18 Jul 1962 (2 males), 20 Jul 1962 (5
males), 4200' [=1280 m.], J. F. G. Clarke (USNM). Klamath
Co.: "61" [Summit Lake, 26 Sep 1871] (1 male), Walsingham
(USNM). **Utah.**- Utah Co.: Provo, 9 Jul 1909 (1 male), 12 Jul
1909 (1 male), T. Spalding (USNM). **Washington.**- Clark Co.:
Booneville, 14 Jul 1931 (1 female), J. F. G. Clarke (USNM).
Klickitat Co.: Satus Pass, 3 Aug. 1962 (3 males), J. F. G.
Clarke (USNM). Whitman Co.: Pullman, 11 Jul 1930 (1
male), J. F. G. Clarke (USNM). Yakima Co.: 2.5 mi. [=4
km.] W. Ft. Simcoe, 31 Jul 1962 (2 males, 6 females), J. F.
G. Clarke (USNM). **Canada.**- **British Columbia.**- Langford,
[Vancouver Is.], 27 Jun 1961 (1 male, 1 female), 30 Jun 1961
(3 males), D. Evans (CNC); Little Malahat Mt., Vancouver
Is., 24 Jun 1926 (1 male, 1 female), J. F. G. Clarke (USNM);
Maple Bay, Vancouver Is., 12 Jul 1933 (1 male), J. H.
McDunnough (CNC); Saanichton, [Vancouver Is.], 4 Jul 1925
(1 male), J. G. Colville (USNM); Vernon, 13 Jul 1927 (1
male), E. P. Venables (CNC); Victoria, [Vancouver Is.], 21

Jun 29 (1 male), J. F. G. Clarke (USNM); 24 Jun 1922 (2
males) W. Downes (CNC); 21 Jul 1920 (1 male), E. H.
Blackmore (LACM); [no date](1 male)(LACM); 14 Jun 1924
(1 male), 19 Jun 1920 (1 male), 21 Jul 1920 (1 male), E. H.
Blackmore (USNM); 16 Jun 1903 (1 male) (USNM); 21 June
1929 (1 male), J. F. G. Clarke (USNM); 17 June 1923 (2
males), 20 Jun 1923 (1 male), 24 Jun 1923 (2 males), W. R.
Carter (USNM).

Distribution (Fig. 7).- Montana to British Columbia,to
southern California and western Nevada; Utah.

Flight period.- June to early September, with most dates
in June-July.

Hosts.- Unknown.

Biology.- Unknown. Adults have been collected in the
vicinity of **Arctostaphylos** (Ericaceae) (Clarke, pers. comm.)
but it is not known whether this represents flower visitation
(although this is probable) or whether the moths were con-
gregating around the host plant.

Remarks.- The Walsingham specimens of 1871 could all
be thought of as syntypes but I have chosen only those noted
in his original description where he oddly noted only three
specimens from near Mt. Shasta and two from near San
Franscisco.

This species is the largest on average of the four species
of the species-group and is distinctive in the full extension
of the basal fascia across the wing in most specimens and in
having only three white marks on the apex. The genitalia
are virtually the same as the other three species but the
female genitalia differ in having the apophyses shorter than
in the others with known females. A few G. bifasciata
specimens have one of the apical forewing costal white
marks reduced, producing the appearance of only two apical
marks instead of three, but most specimens are uniform with
three marks. The species appears closest to the following
new species.

Glyphipterix hypenantia Heppner, new species
(Figs. 8, 78, 152-153, 222-223)

A smaller species than G. **bifasciata**, with four white
marks near the apex of the forewing and the basal marks as
a costal bar and a dorsal margin crescent.

Male (Fig. 78).- 5.6-5.8 mm. forewing length. Head:
fuscous mixed with buff, fine line of white along lateral eye

margin; labial palpus white with fuscous at base of apical
segment dorsally, venter with basal segment white, 2nd segment fuscous basally and white apically, and apical segment
similar, with scale tufts long from 2nd segment; antenna
dorsally fuscous. **Thorax:** fuscous; patagia fuscous; venter
mostly white; legs fuscous with white at joints. **Forewing:**
ground color fuscous with a few brown scales intermixed
toward apex; basal 1/4 with small white crescent on dorsal
margin with tip pointed to oblique white bar on costal margin at 1/3 wing length from base; midwing white fascia interrupted near middle by silver spot; 4 white spots on costal
margin on apical 1/3 of wing, each with a spot of silver
mesally; apical 1/4 of wing with several black scale striae
longitudinally in field of white-buff scales; silver spot at
falcate indentation; tornus with 3-4 silver spots surrounded
by black, with another silver spot slightly basad of black
area; fringe fuscous, white distally, with white at falcate
indentation; venter fuscous with white marks faintly repeated
along costal margin and termen. **Hindwing:** fuscous; fringe
fuscous; venter light fuscous with two faint white spots on
apex. **Abdomen:** fuscous with silver-white scale row on
posterior of each segment; venter similar but mostly white.
Genitalia (Fig. 152): tuba analis large, long; teguman split;
vinculum rounded, convex; saccus long and thin, somewhat
bulbous on distal end, extended as Y-shaped extension to
valval bases through vinculum; valva very elongate, setaceous, with long, thin apical point; short tubular anellus;
aedeagus (Fig. 153) long and thin, 3/4 valval length, with
spicule collar on tip; phallobase short, narrower, cornutus a
short tubule; vesica with spicules; ductus ejaculatorius with
hood near aedeagus (3 preparations examined).

Female.- 4.5 mm. forewing length. Similar to male.
Genitalia (Fig. 223): ovipositor long; papilla analis small,
setaceous; apophyses long, thin, posterior pair almost twice
length of anterior pair minus the basal fork; ostium bursae
(Fig. 222; not fully extended) a membranous, broad funnel
ending in a sclerotized cup; ductus bursae very thin, sclerotized, long; ductus seminalis emergent from ductus bursae
junction with bursa; corpus bursae small, oval; signum absent
(1 preparation examined).

Types.- Holotype male: 1 mi. [=1.6 km.] SE Bartle, Sis-
kiyou Co., California, 11-14 Jun 1974, J. A. Powell (UCB).
Paratypes (17 males, 1 female): **California.-** same data as
holotype (9 males) (UCB), 26 Jul 1974 (8 males), J. A. Powell
(UCB); 20 Jul 1966 (1 female), P. Rude (UCB). (Holotype to
CAS; paratypes to BMNH, CNC, JBH, and USNM).

Distribution (Fig. 8).- Known only from northern California.
Flight period.- June to July.
Hosts.- Unknown.
Biology.- Unknown.
Remarks.- **Glyphipterix hypenantia** is distinguished from G. **bifasciata** by the 4 marks near the apex and the basal two white marks. **Glyphipterix unifasciata** Walsingham is much smaller and does not have the basal white marks on the costal margin, which **Glyphipterix yosemitella**, new species, also lacks. As noted previously, G. **hypenantia** is most closely related to G. **bifasciata**.

The specific name is derived from the Greek for "opposite," referring to the opposed basal wing markings.

Glyphipterix yosemitella Heppner, new species
(Figs. 8, 82, 160–161)

This species is distinguished by the apical 4 white marks of the costal margin, the lack of a basal white mark on the dorsal margin, and the lack of a distinct striae field on the apical quarter of the forewing.

Male (Fig. 82).- 5.4 mm. forewing length. **Head:** fuscous, some buff on lateral eye margin; labial palpus dorsally white–buff and fuscous, venter with basal segment white, 2nd segment alternating white and fuscous, and apical segment alternating fuscous and white–buff, long scale tufts; antenna fuscous dorsally. **Thorax:** fuscous; patagia fuscous; venter fuscous and white; legs fuscous with white at joints. **Forewing:** fuscous ground color, overlaid with brown scales only on apical 1/3 with some buff in reduced tornal striae field where striae are indistinct fuscous; basal 1/4 with thin white oblique bar from dorsal margin to CuP fold; somewhat angulate, oblique white fascia across wing before middle; apical 1/3 of wing with 4 white bars on costal margin, each mesally with a silver spot, longer silver spot mesad of white spot 3rd from apex; silver spot at falcate indentation; tornus with 3–4 silver spots surrounded by black; silver spot and white spot basad of tornus and not surrounded by black and another silver spot mesad of basal silver spot of tornus; fringe fuscous, white distally with all white at falcate indentation; venter fuscous with white marks faintly repeated except more distinct apical white marks, white on anal field. **Hindwing:** fuscous; fringe fuscous; venter fuscous with white on apex. **Abdomen:** fuscous with silver-white scales on posterior of each segment; venter similar but mostly white. Genitalia (Fig. 160): tuba analis long; tegumen split; vinculum broad, convex; saccus very long, thin, with bulbous tip; valva elongate, narrow, setaceous, with projected thin point; short tubular anellus; aedeagus (Fig. 161) long (3/4 valval length) very thin, with long apical spicule collar; cornutus a short tubule; vesica with spicules; phallobase narrower than aedeagus; ductus ejaculatorius with hood near aedeagus (1 preparation examined; damaged in mounting).

Female.- Unknown

Type.- Holotype male: 4 mi. [6.4 km.] S. Mather, Tuolumne Co., California, 12 Jun 1961, J. A. Powell (UCB). (Holotype to CAS).

Distribution (Fig. 8).- Known only from the type locality.

Flight period.- June.

Hosts.- Unknown.

Biology.- Unknown.

Remarks.- **Glyphipterix yosemitella** is closely related to G. unifasciata and appears to be very similar superficially. It is larger than most G. unifasciata, however, and the genitalia are relatively much larger since they are as large or larger than those of G. bifasciata, which generally has a greater wing span than any of the species in the group. The aedeagus of G. yosemitella also is as thin as in G. unifasciata but longer and with a longer spicule collar apically. Glyphipterix yosemitella is generally darker than the other species of the group. Inasmuch as all the species of the group are so close genitalically, the differences evident in G. yosemitella indicate that this Sierra Nevada population represents a montane species distinct from G. unifasciata. The latter species also has an apical white bar on the hind wings and white to buff densely around a silver fascia on the forewing apical 1/4, both characters not found in G. yosemitella.

The specific name is derived from the Yosemite Valley, the type locality.

Glyphipterix unifasciata Walsingham
(Figs. 8, 76–77, 154–155, 224–225)

Glyphipteryx [sic] **unifasciata** Walsingham, 1881:322;
Riley, 1891:104; Dyar, 1900:84; [1903]:493;
Kearfott, 1903:108; Meyrick, 1913b:45; 1914c:32;

Barnes & McDunnough, 1917:182; McDunnough, 1939:84.
Glyphipterix unifasciata.- Heppner, 1982a:54; 1983a: 26.

The smallest species of the species-group is close to **Glyphipterix yosemitella** but has a white bar on the hind wing apex and a long silver bar apically on the forewing surrounded by buff.

Male (Fig. 76).- 3.7-4.8 mm. forewing length.   **Head:** fuscous with buff lateral eye margin; labial palpus white and fuscous dorsally, venter with basal segment white, 2nd segment with 2 alternating tufts of fuscous and white, and apical segment with fuscous and white; antenna fuscous dorsally.   **Thorax:** fuscous; patagia fuscous; venter fuscous, with some white; legs fuscous with white at joints.   **Forewing:** fuscous ground color, overlaid with brown scales only on apical 1/3 of wing; white oblique bar at basal 1/4 on dorsal margin to CuP fold, as wide as midwing fascia; somewhat angulate white fascia across wing before midwing, sometimes nearly divided in middle; apical 1/3 of costa with 4 white bars each with silver mesally, with white bar 3rd from apex usually with silver mesally extended almost to tornal spots; apex with a black crescent; a silver spot at falcate indentation; tornus with 3 silver spots surrounded by black; a white spot basad of tornus; apical field densely buff, often white, surrounding the long silver fascia; fringe fuscous, white distally with all white at falcate indentation; venter fuscous with white dorsal markings faintly repeated, more distinct along costa and apex.   **Hindwing:** narrow; fuscous with white subapical fascia; fringe fuscous, white distally, with mostly white on tornus; venter fuscous with white marks repeated.   **Abdomen:** fuscous with silver-white scales on posterior of each segment; venter similar but mostly white.   **Genitalia (Fig. 154):** tuba analis long; tegumen split; vinculum broad, convex; saccus long, narrow; valva very elongate, narrow, setaceous, with projecting thin apex; short tubular anellus; aedeagus (Fig. 155) elongate (2/3 valval length), thin, with short apical spicule collar; phallobase narrower than rest of aedeagus; cornutus a short tubule; vesica with spicules; ductus ejaculatorius with hood near aedeagus (5 preparations examined).

Female (Fig. 77).- 4.2 mm. forewing length. Similar to male, hindwing white fascia more distinct. **Genitalia (Fig. 225):**   ovipositor long; papilla analis long, setaceous;

apophyses long and thin, posterior pair much longer than anterior pair; ostium bursae (Fig. 224) a long narrow membranous funnel ending in a small sclerotized cup; ductus bursae very long, thin, sclerotized at bursal end; ductus seminalis from ductus bursae and bursa junction; corpus bursae small, oval; signum absent (1 preparation examined).

Types.- Lectotype male by present designation: Dry Creek, Sonoma Co., California, 20-21 May 1871, Walsingham 92011 (BMNH). (Labelled additionally: Walsingham, Collection, 1910-427; TYPE male descr., fig. d [Durrant label]; LECTOTYPE male, **Glyphipteryx unifasciata** Wlsm., By Heppner '76; B.M. male, Genitalia Slide, No. 20229). Paralectotype (1 male): same data as lectotype, Walsingham 92012 (BMNH).

Additional specimens (14 males, 1 female).- **California.**- Contra Costa Co.: Pleasant Hill, May-Aug 1960 (1 male), W. E. Ferguson (UCB). Lake Co.: Scott Valley, 17-19 Jun 1871 (1 male), Walsingham 92013 (BMNH). Marin Co.: Novato, 6 May 1962 (4 males, 1 female), D. C. Remtz (CAS); West Novato, 23 Apr 1961 (4 males), 12 May 1961 (1 male), D. C. Rentz (CAS). Mendocino Co.: 7 mi. [=11.2 km.] SE. Yorkville (1 male), J. A. Powell (UCB). Monterey Co.: Monterey, 23 May 1964 (1 male), P. Rude (UCB). Nevada Co.: 2 mi. [=3.2 km.] S. Grass Valley, 3 Jul 1967 (1 male), J. A. Powell (UCB).

Distribution (Fig. 8).- Known only from the northern California Coast Range and east to the Sierra Nevada foothills.

Flight period.- April to July (the reference to August as "May to August", may be erroneous).

Hosts.- Unknown.

Biology.- Unknown.

Remarks.- On the average this is the smallest species of the group and is distinctive in the white subapical fascia of the hindwings. The forewings are also distinct in having 3 tornal silver spots and a dense buff area around a long silver fascia from the 3rd costal white mark from the apex. The above markings also distinguish the species from the very similar **Glyphipterix yosemitella**, especially in regard to the much darker hindwings of the latter species. **Glyphipterix unifasciata** also has a narrower hindwing in relation to equally sized dwarf specimens of G. **bifasciata** and G. **hypenantia**. The genitalia are virtually identical to G. **yosemitella** except for their generally smaller size and the shorter aedeagus with a shorter apical spicule collar.

Thus far, G. unifasciata is uncommon in collections, although present in a populated and relatively well-collected area of California.

The haworthana species-group

Forewing elongate-oblong, with falcate indentation on termen below rounded apex. Hindwing somewhat oblong or more pointed, termen reduced. Labial palpus long, very dorso-ventrally flattened on apical segment, relatively smooth-scaled, apical segment subequal to long 2nd segment, or 2nd segment not so long. Eye moderate. Abdomen without coremata. Valva elongate, with or without apical point; tegumen entire dorsally; tuba analis relatively short; saccus long, narrow or reduced, no rodlike extension in vinculum to valval bases; aedeagus without phallobase. Ovipositor long; papilla analis strongly sclerotized with cutting edges; apophyses very stout, long; ostium bursae a funnel with sclerotized anterior cup or without any sclerotization; ductus bursae sclerotized or membranous, thin or moderately wide, short; corpus bursae oval with small anterior accessory bursa; signum present as small spicule patches.

In North America north of Mexico this species-group contains two species, the Holarctic Glyphipterix haworthana (Stephens) and the new species described hereafter. It appears that several other Palearctic species related to G. haworthana also belong in this species-group, due primarily to similarities in the genitalia. This group is related to the bifasciata species-group as evident by wing maculation and some genitalic features, in particular the thin sclerotized ductus bursae and the sclerotized anterior cup of G. haworthana. There appear to be close relationships to the californiae and montisella species-groups of the Nearctic, to be discussed subsequently, and to the European bergstraesserella species-group (based on Glyphipterix bergstraesserella (Fabricius)).

Glyphipterix haworthana (Stephens)
(Figs. 9, 84-85, 185-186, 250-251)

Heribeia haworthella Stephens, 1829:207, nom. nud;
    Rennie, 1832:202, nom. nud.; Westwood, 1854:
    194.

Heribeia haworthana Stephens, 1834:262; Wood, 1837:
194; Westwood, 1845:211; 1854:194.
Oecophora zonella Zetterstedt, [1839]:1009.
Aechmia zonella.- Standfuss, 1849:21.
Aechmia haworthella.- Herrich-Schäffer, 1854:93;
Walker, 1864:842.
Glyphipteryx [sic] haworthana.- Stainton, 1854b:175;
1855:4564; 1859:364; 1870:232; Heinemann, 1870:
396; Wock, 1871:309; Bang-Haas, 1875:35; Chambers, 1875b:293; Millière, [1876]:347; Wocke,
[1876]:396; Zeller, 1877:401; Frey, 1880:379;
Hartmann, 1880:93; Chambers, 1881:291; Snellen,
1882:749; Sorhagen, 1885:95; Meyrick, 1895:704;
Reutti, 1898:181; Malloch, 1901:186; Rebel, 1901:
130; Spuler, 1910:298; Meyrick, 1913b:43; 1914c:
30; Hering, 1927:433; Meyrick, 1928:710; Amsel,
1930:116; Schütze, 1931:18; Amsel, 1932:20; Hering, 1932:177; Eckstein, 1933:109; Morley & Rait-
Smith, 1933:178; Sterneck & Zimmermann, 1933:
79; Pierce & Metcalfe, 1935:42; Le Marchand,
1937a:193; Heslop, 1945:27; Kloet & Hincks,
1945:132; Thompson, 1946:262; Lhomme, 1948:
504; Amsel, 1949:88; Wörz, 1954:85; Klimesch,
1961:724; Bleszynski, et al., 1965:413; Chinery,
1972:pl.21,f.18; Hannemann & Urbahn, 1974:320.
Glyphipteryx [sic] haworthella.- Morris, 1872:136;
Morris & Kirby, 1896:136; Porritt, 1904:152.
Glyphipteryx [sic] howarthana Jordan, 1886:154, missp.
Glyphipterix haworthana.- Ford, 1949:127; 1954:99;
Krogerus, et al., 1971:17; Bradley, 1972:12;
Goater, 1974:52; Heath, 1976:102; Karsholt &
Nielson, 1976:25; Leraut, 1980:82; Heppner,
1982a:50; 1983a:26.

This species is easily distinguished from the two similar
North American species by the male genitalia, while also
being much larger than the other two species.

Male (Fig. 84).- 4.8-5.5 mm. forewing length. Head:
fuscous, with lateral white eye margin; labial palpus dorsally
white, venter smooth-scaled, with basal segment white, 2nd
segment twice length basal segment, with fuscous as linear
central line, white laterally, and apical segment similar,
subequal to 2nd segment; antenna dorsally fuscous. Thorax:
fuscous; patagia fuscous; venter fuscous; legs fuscous with

white at joints. **Forewing:** ground color fuscous with apical 1/3 of wing overlaid with brown scales; dorsal margin with large white crescent mark at midwing extended to cell; apical half of costal mark at midwing extended to cell; apical half of costal margin with 5 white marks, each mesad with a silver spot, with midwing mark having a longer oblique silver bar and 4th mark from apex with a convex silver fascia usually extended to tornal white spot without interruption; apex with black crescent and silver spot at falcate indentation; fringe dark fuscous, white distally and all white at falcate indentation; venter fuscous with white marks faintly repeated except distinct apically. **Hindwing:** fuscous; fringe fuscous on apex and tornus, then white from mid-dorsal margin to anal angle; venter fuscous with some pale scaling on anal field and apex. **Abdomen:** fuscous with silvery-white scales on posterior of each segment; venter similar but mostly white. **Genitalia** (Fig. 185): tuba analis well-developed but somewhat sclerotized posteriorly, short; tegumen stout, entire, sometimes with interior projection dorsally; vinculum rounded, convex; saccus long, narrow; valva elongate, setaceous, with curved point on dorsal margin apex; short tubular anellus; aedeagus (Fig. 186) long (subequal to valva), narrow, with long apical spicule collar; phallobase absent; cornutus a moderate tubule; vesica with spicules; ductus ejaculatorius with campanulate hood near aedeagus, with ductus entering aedeagus at anterior end (3 preparations examined).

Female (Fig. 85).- 5.1-6.0 mm. forewing length. Similar to male. Genitalia (Fig. 251): ovipositor long; papilla analis strongly sclerotized, with pointed cutting edges; apophyses long, very stout, with blunt anterior ends, posterior pair longer than anterior pair; ostium bursae (Fig. 250) a wide, short, membranous funnel, with an anterior sclerotized cup; ductus bursae thin, short, sclerotized; ductus seminalis emergent from bursa near ductus bursae entrance; corpus bursae elongate-ovate, with small anterior accessory bursa; signum as two small spicule patches on opposite sides of bursa near middle (3 preparations examined).

Types.- Lectotype male, by present designation: [England] (BMNH). (Labelled as follows: Stephens Coll., **Glyphipteryx** [sic] **haworthana,** c, named by Steph.; LECTOTYPE male, **Glyphipterix haworthana** Steph., By Heppner '76; B.M. male, Genitalia Slide, No. 20227). Paralectotypes (3 males, 4 females): Stephens Coll., a, e, & f (3 males), and b, d, g, and h (4 females). (Specimen "g" is labelled "**haworthella** mihi" but since the abdomen is missing, specimen "c" is chosen as lectotype).

Additional specimens (Nearctic only) (87 males, 78 females).- **Maine.**- Penobscot Co.: Enfield, 20 May (1 male), L. P. Grey (LACM); Passadumkeag, 28 May (1 male), L. P. Grey (LACM). **Canada.**- **Manitoba.**- Cartwright Lab, [Cartwright], 26 Jun 1955 (1 female), E. F. Cashman (CNC). **Northwest Territories.**- Aklavik, 25 Jun 1956 (1 female), E. F. Cashman (CNC). **Nova Scotia.**- Bog E. end Indian Lake, Halifax watershed, 25 May 1959 (1 male, 1 female), D. C. Ferguson (NSM).   **Ontario.**- Mer Bleue, Ottawa, 5 May 1931 (2 males), J. H. McDunnough (CNC); 5 May 1931 (5 males, 2 females), G. S. Walley (CNC); 5-15 May 1965 (2 females), K. Sattler (ZSBM); 13 May 1932 (1 male, USNM; 4 males, 1 female, CNC), W. J. Brown; 13 May 1932 (2 females, USNM; 14 males, 1 female, CNC), G. S. Walley; 13 May 1932 (3 males, USNM; 27 males, 8 females, CNC), J. H. McDunnough; 14 May 1905 (2 males), C. H. Young (USNM); 14 May 1904 (2 males, 3 females), C. H. Young (CNC); 14 may 1965 (4 females, ZSBM; female, CNC), K. Sattler; 15 May 1965 (11 males, 26 females, ZSBM; 1 male, 4 females, CNC), K. Sattler; 16 May 1922 (3 males, 4 females), J. H. McDunnough (CNC); 16 May 1940 (2 males), T. N. Freeman (CNC); 27 May 1905 (1 male, 5 females), C. H. Young (CNC); 25 May 1923 (3 males, 9 females), J. H. McDunnough (CNC); 7 Jun 1923 (1 male, 1 female), R. Ozburn; Smoky Falls, Mattagami River, 13 Jun 1934 (1 male, 1 female), 18 Jun 1934 (1 male), G. S. Walley (CNC).

Distribution (Fig. 9).- Nova Scotia to Manitoba and Northwest Territories; Europe.

Flight period.- May to June (Nova Scotia to Ontario); late June (Manitoba and Northwest Territories); late April to May (Europe).

Hosts.- **Eriophorum angustifolium** Roth, **E. gracile** Koch, and **E. lanatum** [unknown name; ?= **E. latifolium** Hoppe] (Cyperaceae).

Biology.- The larvae bore in the seed heads of **Eriophorum** species, where they overwinter and emerge the next spring.

Remarks.- This is the only known Holarctic species. The European and North American populations are not significantly different.   North American males have a more pronounced interior projection of the tegumen (Fig. 185) but this projection is present, though reduced, in populations from England as well. Some specimens, especially the two from Manitoba and Northwest Territories, have the dorsal

margin crescent very narrow, as do many specimens from
Europe, while most other North American specimens have
this mark somewhat broad. The European specimens tend to
be more lustrous fuscous on the forewings than do the North
American specimens.

Glyphipterix haworthana has several related species in
Europe but in North American the forewing maculation is
similar only to Glyphipterix sistes, new species, and
Diploschizia impigritella (Clemens) and Diploschizia kimballi
Heppner. These species are all much smaller than G. haworthana and are easily distinguished by the male genitalia.
In the past G. haworthana from North America (as also G.
sistes, n. sp.) has been misidentified as "G." imprigritella,
since it was not examined genitalically and was thought to
be only a giant example of the latter species.

In the females, both G. haworthana and G. sistes, new
species, have similar genitalia, both with strongly sclerotized
ovipositors evidently capable of piercing oviposition.

Glyphipterix sistes Heppner, new species

The 5-spotted costal margin and dorsal crescent of the
forewing and small size will distinguish this species from all
North American species except for two which belong in
Diploschizia characterized by lacking vein M3 of the hindwings. The species is known from the Pacific and Atlantic
Coasts of southern Canada and the northern United States.
The nominate subspecies has a thin dorsal margin crescent
and has more of a bronze-fuscous iridescence on the forewing, while the eastern populations appear grayer and have a
broader forewing crescent.

Glyphipterix sistes sistes Heppner, new subspecies
(Figs. 9, 86-87, 187-188, 252-253)

Male (Fig. 86).- 3.4-3.8 mm. forewing length. Head:
gray fuscous with white lateral eye margin; labial palpus
relatively long but 2nd segment not very long, subequal to
apical segment; dorsally white, ventrally with basal segment
white, 2nd segment with 2 alternating bands of fuscous and
white, somewhat roughened, and apical segment similar to
2nd segment but smooth-scaled; antenna fuscous dorsally.
Thorax: fuscous; patagia fuscous; venter mostly white, some
fuscous; legs with femur and tibia white, tarsal segments
fuscous with white at joints. Forewing: lustrous bronze-

fuscous, somewhat grayish to olive overlayer, with dark fuscous emarginate on apical white and silver markings; large narrow white dorsal margin crescent at midwing; distal half of costal margin with 5 white marks, with 4th and 3rd from apex having a mesal silver spot; apex with spot of black; small silver spot at falcate indentation and mid-apically mesad of indentation; tornus with 2 silver spots partially surrounded by black; white spots at basal end of tornus, followed mesad by a silver bar; fringe fuscous, distally white at apex, all white at falcate indentation; venter fuscous, with white spots repeated near apex, silvery-white on anal field. **Hindwing:** very narrow with apex pointed fuscous; fringe fuscous, long along costal margin; venter fuscous with white along costal margin at apex. **Abdomen:** fuscous with brown scales on posterior of each segment; venter mostly silvery-white, some fuscous on anterior of each segment. **Genitalia** (Fig. 187): tuba analis somewhat short; tegumen stout laterally, thinner dorsally; vinculum rounded, convex; saccus reduced to short stub; valva elongate oblong, thick, with rounded apex, setaceous, short tubular anellus; aedeagus (Fig. 188) short (2/3 valval length), thin, without phallobase, short spicule collar; cornutus a long tubule (1/2 aedeagus length); vesica with spicules; ductus ejaculatorius with campanulate hood approximate to aedeagus (7 preparations examined).

Female (Fig. 87).- 3.6–4.1 mm. forewing length. Similar to male but with brown scales added to forewing layer over ground color. **Genitalia** (Fig. 253): ovipositor long; papilla analis strongly sclerotized with acute cutting edges; apophyses very stout, long with blunt anterior ends, posterior pair 1/2 longer than anterior pair; ostium bursae (Fig. 252) a wide membranous funnel; ductus bursae wide, membranous, short; ductus seminalis from near middle of ductus bursae length; corpus bursae ovate, with small accessory bursa anterior; signum with 2 spicule patches on opposite sides of bursa (3 preparations examined).

Types.- Holotype female: Duncan, Vancouver Is., British Columbia, 31 Jul 1925, A. W. Hanham (USNM). Paratypes (30 males, 15 females): **Alaska.**- Ketchikan, 26 Apr 1916 (1 male), B. P. Clark (USNM); Matanuska, 21 May 1945 (male), J. C. Chamberlin (CNC). **California.**- El Dorado Co.: Blodgett Forest, 13 mi. [=20.8 km] E. Georgetown, 5 Jun 1975 (1 male), P. Rude (UCB). Humboldt Co.: Fieldbrook, 26 May 1903 (1 male), H. S. Barber (USNM). Mendocino Co.: Navarro River, 3 mi. [=4.8 km] SE. Dimmick St. Park, 25-30 Jul 1971 (1 male), P. Rude (UCB). Placer Co.: Bear Valley (east end), 1 Jun 1964 (1 male), P. H. Arnaud, Jr. (CAS). Tuolumne–Alpine Co.: Utica Reservoir, 30 Jun-4 Jul 1971 (1 male, 6 females), 6800' [=2070 m.] ex Malaise trap, D Veirs (UCB). **Oregon.**- Klamath Co.: Eagle Ridge, Klamath Lake, 19 May 1924 (1 female), C. L. Fox (CAS). **Canada.**- British **Columbia.**- Duncan, Vancouver Is., 29 Jul 1925 (3 males), 31 Jul 1925 (2 males), 3 Aug 1925 (2 males, 2 females), 6 Aug 1925 (1 male, 3 females), 10 Aug 1925 (3 males), 11 Aug 1925 (3 males, 1 female), 17 Aug 1925 (4 males, 1 female), 20 Aug 1924 (2 males), A. W. Hanham (USNM); Quamichan Lake, Vancouver Is., 6 Aug 1923 (1 male), 10 Aug 1923 (1 male), A. W. Hanham (USNM); Squamish, Diamond Head Trail, 26 Aug (1 male, 1 female), 3200' [=975 m.] W. R. M. Mason (CNC).

Distribution (Fig. 9).- Southern Alaska to California.

Flight period.- April to May (Alaska); July to August (British Columbia); May to July (Oregon and California).

Hosts.- Unknown.

Biology.- Unknown.

Remarks.- The genitalia of this species are distinct from any other North American species of **Glyphipterix.** The western subspecies is confined to the West Coast as far as is known but since even Washington state is not represented among the available specimens, more collecting may locate populations further into the interior. Western specimens are lustrous bronze fuscous in forewing coloration while the eastern subspecies is grayer in appearance. The forewing dorsal crescent is almost always rather narrow in the western subspecies but some specimens from California approach the extent of thickening of this mark as found in the eastern subspecies.

The specific name is derived from Greek meaning to "move to and fro", which the adults do when perching on plants.

Glyphipterix sistes viridimontis Heppner, new subspecies
(Figs. 9, 88-89)

This very allopatric subspecies from the Northeast is distinguished by the broader, more quadrate base of the forewing crescent mark of the dorsal margin and by the overall grayer appearance relative to the nominate subspecies.

Male (Fig. 88).- 3.6-3.8 mm. forewing length. Similar to the nominate subspecies with the following differences: instead of a bronze iridescence on fuscous areas of head, thorax, and forewing, these areas are gray-fuscous, with black basad of which dorsal margin crescent mark and a black spot at center of apical 1/4 of wing; dorsal margin white crescent much wider and usually nearly quadrate near margin. **Genitalia:** identical to nominate subspecies (2 preparations examined).

Female (Fig. 89).- 3.5-4.0 mm. forewing length. Similar to male.

Genitalia: identical to nominate subspecies (4 preparations examined).

Types.- Holotype female: Chittenden Brook Cpgd., Green Mountains, Rutland Co., Vermont, 18 Aug 1973, on **Solidago** flowers, J. B. Heppner (USNM). Paratypes (8 males, 15 females): **Vermont.**- same data as holotype (JBH). (Paratypes to BMNH, CNC, FSCA, UCB, and USNM).

Additional specimens (1 male, 1 female).- **Maine.**- Hancock Co.: Pretty Marsh, Mt. Desert Is., 25 Aug 1944 (1 male), A. E. Brower (CPK). **Canada.**- **Nova Scotia.**- Queens Co.: White Point Beach, 15 Aug 1935 (1 female), J. H. McDunnough (CNC).

Distribution (Fig. 9).- Collection records are only for Nova Scotia, Maine and Vermont.

Flight period.- August.

Hosts.- Unknown.

Biology.- Unknown, but in Vermont the adults were feeding on flowers of a **Solidago** species (Compositae) next to Chittenden Brook in a montane deciduous forest meadow.

Remarks.- This subspecies was already discussed previously in terms of relationship to the nominate subspecies. There are too few specimens of the eastern subspecies to show any significant variation and all appear as the two specimens illustrated except that the forewing dorsal crescent has minor variations in the shape of the base on the dorsal margin.

The subspecific name is derived from Latin for "green mountain" and refers to the Green Mountains of the type locality.

The californiae species-group

Forewing elongate-oblong, with falcate indentation following rounded acute apex. Hindwing pointed, no distinct termen. Labial palpus long, relatively smooth-scaled ventrally, with apical segment longer than 2nd segment, pointed and dorso-ventrally flattened. Eye moderate. Abdomen usually with coremata ventrally near the base of the valvae on the 8th abdominal segment. Valva short, setaceous, very oblong, usually with a saccular point; tegumen entire but thin dorsally; tuba analis long, narrow, with lateral sclerotized ridge each side; saccus very short, narrow without extensions into vinculum. Ovipositor usually long, rarely short; apophyses long and relatively stout or short and thin; papilla analis simple, setaceous; ostium bursae a small membranous funnel with an anterior sclerotized cup; ductus bursae short, thin, sclerotized; ductus seminalis emergent from bursa proximal to ductus bursae; corpus bursae ovate, moderate to relatively large or elongate, without accessory bursa or at most only a very reduced accessory bursa; signum absent.

The species-group comprises 6 species in North America north of Mexico and appears restricted to the Nearctic, with one undescribed species from Nearctic portions of northern Mexico. Only one species has been described in the group until now, **Glyphipterix californiae** Walsingham, and it was initially thought that this species was widespread over western North America. As the various populations were studied it became evident that a species complex was involved including 3 generally allopatric species and 3 sympatric species. The species are all very close, but have various maculation and genitalia details that distinguish them. The group appears to be most closely related to the **montisella** species-group.

Glyphipterix californiae Walsingham
(Figs. 10, 92-93, 175-176, 238-239)

Glyphipteryx [sic] californiae Walsingham, 1881:320;
    Riley, 1891:104; Dyar, 1900:84; [1903]:493;
    Kearfott, 1903:108; Meyrick, 1913b:44; 1914c:30;
    Barnes & McDunnough, 1917:182; Braun, 1925:
    204; McDunnough, 1939:84.
Glyphipterix californiae.- Heppner, 1982a:48; 1983a:
    26.

The species is very similar to the other 6 species of this species-group but is distinguished by the yellow-buff head and the usually extended mesal end of the dorsal marginal forewing white mark recurved along the CuP fold to the wing base, though this is sometimes vestigial.

Male (Fig. 92).- 4.6-6.2 mm. forewing length. Head: yellow-buff, rarely some fuscous mixed in; labial palpus dorsally white, ventrally with basal segment white, 2nd segment with 2 alternating bands of fuscous and white, and apical segment fuscous with white laterally; antenna fuscous dorsally. Thorax: buff, with some fuscous mixed in; patagia buff; venter mostly silvery white; legs silvery white on femora, remainder fuscous with white at joints. Forewing: ground color fuscous with brown-buff overlaid scales on apical half except for fuscous borders to markings; basal 1/4 with trapezoidal white fascia from dorsal margin, to CuP fold and sharply recurved along fold to wing base, with extension of mark as thin line along anal margin; costal margin with 5 white bars in decreasing lengths from basal 1/3 of wing to apex, sometimes with 3rd mark from apex reduced, all with a silver spot on mesal ends, with 4th bar from apex with large silver spot in cell, then extended to white oblique fascia on dorsal margin near midwing; 2 silver spots beyond end of cell; silver spot at falcate indentation and 3 silver spots along tornus, with basal spot having white distally; dark fuscous to near black on apex; fringe fuscous, distally dark fuscous on apex, then white distally to near tornus, with all white at falcate indentation; venter fuscous with dorsal markings faintly repeated except distinct apical marks. Hindwing: fuscous; fringe fuscous; venter fuscous with some pale white as overlaying scales. Abdomen: fuscous with silvery scales on posterior of each segment; venter similar but mostly white; coremata present. Genitalia (Fig. 175): tuba analis long with lateral sclerotized ridge; tegumen entire, narrow; vinculum subrectangular, convex with central anterior margins concave where saccus joins; saccus short, thin; valva short, oblong with broad rounded distal end, setaceous, recurved to short point at saccular margin; short tubular anellus; aedeagus (Fig. 176) long (subequal to tuba analis-saccus distance), thin; cornutus a short tubule; vesica with spicules; short phallobase; ductus ejaculatorius with campanulate hood near aedeagus (3 preparations examined).

Female (Fig. 93).- 5.0-5.9 mm. forewing length. Similar to male. Genitalia (Fig. 239): ovipositor elongate; papilla analis small, setaceous; apophyses stout, long, with blunt tips, posterior pair nearly twice length anterior pair anterior to basal fork; ostium bursae (Fig. 238) a small funnel with an anterior sclerotized cup; ductus bursae thin, short, sclerotized; ductus seminalis from bursa proximal to ductus bursae; corpus bursae, moderate; signum absent (2 preparations examined).

Types.- Lectotype male, by present designation: Hatchet Creek, Shasta Co., California, 14-17 Jul 1871, Walsingham 92027 (BMNH). (Labelled additionally: Walsingham, Collection, 1910-427; TYPE male descr. fig[ure]d [Durrant label]; LECTOTYPE male, **Glyphipteryx californiae** Wlsm., By Heppner '76; B.M. male, Genitalia Slide, No. 20232). Paralectotypes (7 males, 2 females): same data as lectotype, Walsingham 92028-31 and 92033-34 (6 males), 92032 (1 female) (BMNH); (1 female) (MCZ, type 14991).

Additional specimens (24 males, 7 females).- **California.**- Colusa Co.: "25" [Phip's Place, 26 Jun 1871] (1 male), Walsingham (BMNH); "26" [Bear Valley, 27 Jun 1871] (1 male), Walsingham (USNM). Shasta Co.: 2.3 mi. [=3.7 km.] SW. Hatchet Mtn. Pass, 15 Jun 1970 (2 males, 1 female), R. M. Brown (CAS); Pit River, 21-26 Jun 1871 (2 males), Walsingham 92035-36 (BMNH). Siskiyou Co.: Mt. Shasta City St. Fish Hatchery, 1 Jul 1970 (5 males, 4 females), P. Rude (UCB). Sonoma Co.: Two Rock, 7 May 1939 (1 male), E. C. Johnston (CNC). Trinity Co.: Big Flat Cpgd., 29 Jun 1974 (9 males, 2 females), 5070' [=1550 m], E. Rogers (UCB); 6 mi. [=9.6 km.] NE Hayfork, 20 May 193 (1 male), J. A. Powell (UCB). [California?], [no date] (1 male), [Walsingham?] (USNM). **Oregon.**- Klamath Co.: "60" [near Crooked Creek, 25 Sep 1871] (1 male), Walsingham (USNM).

Distribution (Fig. 10).- Records are from the southern Cascade Range from southern Oregon to northern California, with extensions into the California Coast Range.

Flight period.- May to July; September (Oregon).

Hosts.- Unknown (probably **Juncus** (Juncaceae)).

Biology.- Unknown.

Remarks.- The buff head coloration is distinctive for this species. The forewing generally has the basal dorsal marginal mark recurved back to the base along the cubital fold, while in the other species of this group it is not recurved but invariably pointed toward the apex of the wing. The genitalia are similar to the other species of the group, except for **Glyphipterix roenastes**, new species, but the saccular tip of the valva is consistently shorter and thinner than in the other species. The undescribed species from Mexico, however, is very similar. In North America north of Mexico, **G. californiae** is closely related to the Great Basin and Rocky Mountain species, **Glyphipterix juncivora**, new species, but closest relationships may be to **Glyphipterix feniseca**, new species.

Glyphipterix feniseca Heppner, new species
(Figs. 10, 96-97, 167-168, 230-231)

This Californian species is similar to **Glyphipterix jun-civora** in having a gray-fuscous head, but has the forewing dorsal margin mark almost Z-shaped and has spatulate anterior apophyses.

Male (Fig. 96).- 5.8-6.0 mm. forewing length. **Head:** gray-fuscous with some buff and buff as lateral eye margin line, some buff on posterior fringe; labial palpus dorsally buff, venter with basal segment white, 2nd segment with 2 alternating bands of fuscous and buff-white, and apical segment fuscous with buff-white laterally; antenna dorsally fuscous. **Thorax:** gray-fuscous; patagia gray-fuscous; venter fuscous and white; legs fuscous and white, with white at joints. **Forewing:** fuscous ground color with brown-buff scales overlaid on apical half of wing except for fuscous emarginations bordering markings; dorsal margin with white mark on basal 1/4 extended along anal margin to base of wing and usually with mesal-distal end extended at cubital fold to give a slanted Z-shaped mark; short white crescent mark beyond midwing on dorsal margin, with silver spot mesad almost convergent with equal but thinner white and silver mark from costa; costal margin with long, curved white crescent before midwing, then 4 more marks to apex, each with a silver spot mesad; 2 silver spots beyond cell; silver spot at falcate indentation; 2 silver spots along tornus and a small white and silver spot at basal end of tornus; apex with black spot; fringe dark fuscous, white distally, with all white at falcate indentation; venter with dorsal white marks faintly repeated except distinct apical marks. **Hindwing:** lustrous fuscous; fringe fuscous; venter lustrous whitish fuscous with small spot of white on apical fringe. **Abdomen:** fuscous with silvery scales on posterior of each segment; venter similar but mostly white; coremata present. **Genitalia** (Fig. 167): tuba analis long, with lateral sclerotized ridge each side; tegumen narrow, thin dorsally; vinculum subrectangular, convex, with concave edge where saccus joins; saccus short, thin; valva short, oblong with large rounded termen, setaceous, recurved to short point at saccular tip; short tubular anellus; aedeagus (Fig. 168) moderate (subequal to valval length), thin; cornutus a short tubule, somewhat sinuate; vesica with spicules; phallobase short;

ductus ejaculatorius with campanulate hood near aedeagus (3 preparations examined).

Female (Fig. 97).- 5.5-6.5 mm. forewing length. Similar to male. Genitalia (Fig. 231): ovipositor long, papilla analis small, setaceous; apophyses long, with posterior pair twice length anterior pair, anterior to basal fork, with anterior pair stouter and with spatulate anterior ends; ostium bursae (Fig. 230) a small membranous funnel with an anterior sclerotized cup; ductus bursae short, thin, sclerotized; ductus seminalis emergent from bursa proximal to ductus bursae; corpus bursae moderate, ovate; signum absent (1 preparation examined).

Types.- Holotype male: North Beach, Point Reyes Natl. Seashore, Marin Co., California, 11 May 1974, G. Bunker (UC 185077) (UCB). Paratypes (8 males, 2 females): California.- Alameda Co.: 12 Apr 1908 (1 male), G. R. Pilate (ANSP). Contra Costa Co.: Antioch, 15 May 1958 (1 male), J. A. Powell (UCB); Lafayette, 9 Apr 1964 (5 males), R. M. Brown (CAS). Marin Co.: Alpine Lake, 6 June 1957 (1 female), 25 Apr 1958 (1 male, 1 female), J. A. Powell (UCB). (Holotype to CAS; paratype to USNM).

Distribution (Fig. 10).- Known only from the San Francisco Bay area.

Flight period.- April to early June.

Hosts.- Unknown (probably **Juncus** (Juncaceae)).

Biology.- Unknown.

Remarks.- Although close to both **Glyphipterix californiae** and G. **juncivora,** new species, it appears to be most closely related to the former species. The spatulate ends of the anterior apophyses of the female genitalia are the most distinctive feature of the genitalia. In the forewing maculation, the slanted Z-shaped dorsal marginal mark is most distinctive.

The specific name is derived from Latin for "harvester."

Glyphipterix juncivora Heppner, new species
(Figs. 11, 94-95, 165-166, 228-229)

This species occurs in the Great Basin region and the adjoining Rocky Mountains. It is distinguished by the lack of a recurved portion of the dorsal marginal basal forewing mark, the gray-fuscous head, and the longer valval saccular point.

Male (Fig. 94).- 5.3-6.0 mm. forewing length. Head: gray fuscous, with buff-white lateral eye margin; labial palpus dorsally buff-white, venter with basal segment white, 2nd segment with 2 alternating bands of white and fuscous, and apical segment with fuscous center and lateral white margin; antenna dorsally fuscous. Thorax: gray-fuscous; patagia gray-fuscous; venter white and some fuscous; legs fuscous with white at joints. Forewing: gray-fuscous ground color, some buff overlaid scales on apical half of wing; dorsal margin with large basal 1/4 quadrate white mark and a smaller wedge-like white mark near midwing, the latter with a mesad silver spot almost convergent with a silver spot from a costal margin white mark; costal margin with 5 white marks from basal 1/3 of wing to apex in decreasing lengths, with basal mark longest and curved; black spot at end of cell and 2 silver spots somewhat beyond; silver spot at falcate indentation; 2 silver spots along tornus; white and silver spot at basal end of tornus; black spot on apex; fringe fuscous, distally white with all white at falcate indentation; venter fuscous with dorsal white marks faintly repeated except distinct apical marks. Hindwing: lustrous fuscous; fringe fuscous; venter lustrous light fuscous. Abdomen: fuscous with silvery scales on posterior of each segment; venter similar but mostly white; coremata present. Genitalia (Fig. 165): tuba analis long with lateral sclerotized ridge each side; tegumen narrow; vinculum subrectangular, convex, with concave edge where saccus joins; saccus short, thin; valva short, very oblong, with rounded setaceous termen, recurved to short saccular point; short tubular anellus; aedeagus (Fig. 166) long (subequal to tuba analis-saccus length), thin; cornutus a short tubule; vesica with spicules; phallobase small; ductus ejaculatorius with campanulate hood near aedeagus (11 preparations examined).

Female (Fig. 95).- 4.8-5.6 mm. forewing length. Similar to male. Genitalia (Fig. 229): ovipositor long; papilla analis small, setaceous; apophyses long with blunt tips, thin, posterior pair twice length anterior pair, anterior to basal fork; ostium bursae (Fig. 228) a small membranous funnel with an anterior sclerotized cup; ductus bursae short, thin, sclerotized; ductus seminalis emergent from bursa proximal to ductus bursae; corpus bursae ovate, small; signum absent (2 preparations examined).

Types.- Holotype male: Lake Hill Cpgd., 7 mi. [=11.2 km.] E. Ephraim, Sanpete Co., Utah, 10 Jul 1976, 8500 [=2590 m.], on Juncus, J. B. Heppner (USNM). Paratypes

(165 males, 32 females): **Arizona.**- Apache Co.: Alpine Divide Cpgd., 4 mi. [=6.4 km] N. Alpine, 13 Jul 1977 (34 males, 3 females), on **Juncus**, T. D. Eichlin & J. B. Heppner (USNM). **Colorado.**- Sprague's, Rocky Mtn. Natl. Park, 3 Aug 1923 (1 female), G. H. & J. L. Sperry (USNM). **New Mexico.**- San Miguel Co.: Beulah, Jul (1 male, 1 female), Cockerell (USNM); Cowles, Sangre de Cristo Mts., 27 Jul 1939 (1 male), A. F. Braun (ANSP). Taos Co.: Taos Ski Area, Sangre de Cristo Mts., 19 Jul 1977 (6 males, 1 female), 9300' [=2380 m] on **Juncus**, J. B. Heppner (USNM). **Utah.**- Cache Co.: 1 Jul 1924 (7 males, 5 females, ANSP; 7 males, USNM; 1 female, CAS), A. F. Braun. Daggett Co.: Navajo Cliffs Cpgd., 6 mi. [=9.6 km.] S. Manila, 7 Jul 1976 (2 males, 2 females), on **Juncus**, J. B. Heppner (JBH). Garfield Co.: Henry Mts., 24 mi [=38.4 km] S. Hanksville, 20 Jul 1968 (6 males, 3 females), 7500' [=2280 m] J. E. H. Martin (CNC). San Juan Co.: Green Mt., La Sal Mts., 13 Jul 1933 (1 male), (AMNH). Sanpete Co.: Lake Hill Cpgd., 7 mi. [=11.2 km.] E. Ephraim, 10 Jul 1976 (22 males, 4 females), 8500' [=2290 m.], on **Juncus**, J. B. Heppner (JBH). Sevier Co.: Bowery Creek Cpgd., Fish Lake, 11 Jul 1976 (78 males, 10 females), 9000' [=2740 m.], on **Juncus**, J. B. Heppner (JBH). **Wyoming.**- Yellowstone R., Yellowstone Natl. Park, 24 Jul 1928 (1 female), J. H. McDunnough (CNC). (Paratypes to BMNH, CDAS, CNC, FSCA, UCB, and USNM).

Additional specimens (4 males).- **Arizona.**- "Arizona", 1882 (3 males), Morrison (Walsingham Coll. 35561, 35563, 35565) (BMNH). **Canada.**- **Alberta.**- Waterton Lakes [Prov. Park], 11 Jul 1923 (1 male), J. H. McDunnough (CNC).

Distribution (Fig. 11).- Alberta to New Mexico, Utah to Arizona.

Flight period.- July to August.

Hosts.- ?**Juncus** sp. (Juncaceae).

Biology.- The adults are collected only in close proximity to the presumed host, one or more species of **Juncus**, along creeks or rivers.

Remarks.- The species is widespread in the Great Basin and Rocky Mountains and is the commonest species of **Glyphipterix** in this region. The single Alberta specimen is disjunct further north and is not designated a paratype. The Arizona specimens collected by Morrison need to be checked again at the British Museum (Natural History) to verify their identification. The species shows variations in details of maculation, in some cases having the inter-marking lighter scaling over the fuscous ground color light enough to appear as additional costal marks, but otherwise the specimens are rather uniform. As in other **Glyphipterix** having fuscous on the forewings, fresh specimens are much darker than flown specimens.

Glyphipterix juncivora is very similar to G. **californiae** but the gray-fuscous head easily distinguishes it.   The species is also similar to G. **feniseca** but this species has a shorter aedeagus, spatulate ends on the anterior apophyses and a slanted Z-shaped dorsal margin white mark instead of the large quadrate mark of G. **juncivora**. **Glyphipterix roenastes**, new species, is superficially most similar to G. **juncivora** but has very different genitalia.

The specific name is derived from Latin for "**Juncus** feeder."

Glyphipterix sierranevadae Heppner, new species
(Figs. 10, 98-99, 169-170, 232-233)

This species from the Sierra Nevada of California has the forewing dorsal margin basal white mark more elongate than the other 6 species of this species-group and the male genitalia are small, with the saccular point of the valva distad enough to reduce the recurved portion of the valval termen.

Male (Fig. 98).- 4.8-6.0 mm. forewing length.   Head: lustrous fuscous, with white lateral line along eye margin; labial palpus dorsally white, venter with basal segment white, 2nd segment with 2 alternating bands of fuscous and white, and apical segment fuscous with lateral edges of white; antenna fuscous dorsally. **Thorax:** lustrous fuscous; patagia fuscous; venter white, some fuscous; legs fuscous with white at joints. **Forewing:** fuscous ground color overlaid with yellow-buff on apical half of wing except for fuscous emarginate lines around markings; basal 1/4 with long white crescent from dorsal margin to CuP fold and toward apex, with a basal white elongation along anal margin; midwing white bar from dorsal margin ending mesad with silver spot almost convergent with similar white and silver bar from costa; costal margin with long white crescent at 1/3 from base, pointed to silver spots at midwing, then margin with usually 4 (sometimes 5) additional white marks to apex, usually each with a small silver spot mesad; dark fuscous spot at end of cell followed by two silver spots; black spot at apex; tornus with 2 silver spots, then a white spot at basal end of tornus; fringe fuscous, pale distally with all white at falcate indentation; venter fuscous with dorsal white marks faintly repeated except apical 3 distinct marks. **Hindwing:** fuscous; fringe fuscous; venter pale fuscous with some white on apical fringe. **Abdomen:** fuscous with silvery scales on posterior of each segment; venter same but mostly white on posterior of each segment; coremata present. **Genitalia** (Fig. 169): tuba analis long with lateral sclerotized ridge; tegumen narrow, thin dorsally; vinculum subrectangular, convex, with concave edge where saccus joins; saccus short, narrow; valva short, very oblong, setaceous, with broad distal end nearly truncate to pointed saccular wide point; short tubular anellus; aedeagus (Fig. 170) short (subequal to valval length), thin; cornutus a short tubule; vesica with spicules; short phallobase; ductus ejaculatorius with campanulate hood near aedeagus (3 preparations examined).

Female (Fig. 99).- 5.6 mm. forewing length. Similar to male. **Genitalia** (Fig. 233): ovipositor long; papilla analis small, setaceous; apophyses long, stout, with posterior pair twice length anterior pair to posterior fork, both pair with blunt tips; ostium bursae (Fig. 232) a deep membranous funnel with an anterior sclerotized cup; ductus bursae short, relatively thin, sclerotized; ductus seminalis emergent from bursa proximal to ductus bursae; corpus bursa large, ovate; signum absent (1 preparation examined).

Types.- Holotype male: Mineral King, Tulare Co., California, 1-7 Jul (USNM). Paratypes (17 males, 1 female): **California.-** Tulare Co.: same data as holotype (11 males, USNM; 1 male, LACM), 24-31 Jul (1 female, USNM); 17 Jul 1913 (1 male, USNM; 1 male, ANSP), 8000' [=2440 m.], A. F. Braun; Rattlesnake Cr., 2 Aug 1915 (3 males) (CU). (Paratypes to BMNH, CNC, UCB).

Distribution (Fig. 10).- Known only from the Sierra Nevada Mountains in Tulare County, California.

Flight period.- July to early August.

Hosts.- Unknown (probably **Juncus** sp (Juncaceae)).

Biology.- Unknown.

Remarks.- This species is close to G. **feniseca** but has the dorsal margin white mark much more elongate. The female genitalia lack the spatulate ends of the anterior apophyses and the ostium bursae is a deeper funnel than in **G. feniseca** or the other species of the group. The male genitalia are distinct in the species-group in having the saccular point of the valva further toward the termen, thus, making the termen not recurved to the saccular point.

The species is named after the Sierra Nevada Mountains of the type locality.

Glyphipterix arizonensis Heppner, new species
(Figs. 11, 100-101, 171-172, 234-235)

This species is distinguished by the somewhat rounded wedge-shaped basal white mark of the forewing dorsal margin. The male genitalia are the most distinctive in the species-group and average smaller.

Male (Fig. 100).- 4.2-5.6 mm. forewing length. **Head:** fuscous, with buff-white line along eye margin; labial palpus dorsally buff-white, venter white on basal segment, 2nd segment with 2 alternating bands of fuscous and white, and apical segment fuscous with lateral edges of white; antenna fuscous dorsally. **Thorax:** fuscous; patagia fuscous; venter fuscous and white; legs fuscous with white at joints. **Forewing:** fuscous ground color, overlaid with paler scales on apical half of wing except emarginate fuscous borders of markings; basal 1/4 of dorsal margin with wedge-shaped white mark not reaching CuP fold, extended along anal margin; costal margin with 5 white bars in decreasing lengths from basal 1/3 to apex of wing, with 4th from apex nearly merging with wide trapezoidal fascia from dorsal margin; some silver in between and silver spot mesad of apical 3 white marks; center of apical 1/4 with 2 silver spots; apex with nearly black spot; silver spot at falcate indentation; 2-3 silver spots along tornus with white and silver spot at basad end of tornus, white distally with all white at falcate indentation; venter fuscous with dorsal white markings faintly repeated except distinct apical marks. **Hindwing:** fuscous; fringe fuscous; venter whitish-fuscous. **Abdomen:** fuscous with silvery scales on posterior of each segment; venter similar but scale rows broadly white; coremata present. **Genitalia** (Fig. 171): tuba analis long with lateral sclerotized ridge each side; tegumen narrow, thin dorsally, vinculum rectangular, narrowing towards saccus, with concave edge where saccus joins; saccus short, narrow; valva very oblong, setaceous, with rounded large distal end, with saccular point broad, distant from valval termen and pointed ventrally; short tubular anellus; aedeagus (Fig. 172) short (subequal to valva), thin; cornutus a short tubule; phallobase short; vesica with spicules; ductus ejaculatorius with campanulate hood near aedeagus (2 preparations examined).

Female (Fig. 101).- 5.3 mm. forewing length. Similar to male. **Genitalia** (Fig. 235): ovipositor long; papilla analis small, setaceous; apophyses long, stout, with blunt tips, posterior pair twice length anterior pair, anterior to posterior fork; ostium bursae (Fig. 234) a deep membranous funnel with an anterior V-shaped sclerotized cup; ductus bursae short, thin, sclerotized; ductus seminalis emergent from bursa copulatrix approximate to ductus bursae; corpus bursae ovate, moderate; signum absent (1 preparation examined).

Types.- Holotype male: Madera Cyn., Santa Rita Mts., Santa Cruz Co., Arizona, 27-30 Jul 1947, L. M. Martin (LACM). Paratypes (5 males, 1 female): **Arizona.-** same data as holotype (3 males) (LACM). "Arizona", 1882 (2 males, 1 female), Morrison, Walsingham Coll. 35564 & 35566 (2 males) and 35562 (1 female) (BMNH). (Paratype to USNM).

Distribution (Fig. 11).- Known only from Arizona.

Flight period.- July.

Hosts.- Unknown.

Biology.- Unknown.

Remarks.- The female genitalia appear closest to G. **sierranevadae** while the wing maculation appears close to G. **juncivora.** The male genitalia are distinctive in having the saccular point ventrad and in their overall small size.

The holotype was chosen from the Madera Canyon specimens inasmuch as this is a more definite locality than the "Arizona" of the Morrison specimens.

The species is named after Arizona and is the same name chosen by Walsingham for the species as the Morrison specimens are labelled; the species was never named in a publication by Walsingham.

Glyphipterix roenastes Heppner, new species
(Figs. 11, 90-91, 173-174, 236-237)

The wing maculation of this species is very similar to **Glyphipterix juncivora** but the white marks of the costal margin are thinner and the genitalia are very distinct.

Male (Fig. 90).- 4.5-5.2 mm. forewing length. Head: fuscous with white eye margins; labial palpus dorsally white, ventrally with basal segment white, 2nd segment with 2 alternating bands of fuscous and white, and apical segment fuscous with white laterally after basal fuscous and white band; antenna fuscous dorsally. **Thorax:** fuscous; patagia fuscous; venter silvery-white; legs fuscous with white at joints. **Forewing:** fuscous ground color with pale buff scaling overlaid on apical half except fuscous borders of markings; dorsal margin with large quadrate white mark on basal 1/4 of wing with white line along anal margin; 2nd white mark on dorsal margin at midwing, widened on margin, extended to end on cell with silver spot mesad almost merging with silver spot of costal white bar; costal margin with 5 narrow white bars from basal 1/3 of wing to apex in decreasing lengths, with silver spot mesad each bar of 4 closest to apex; black spot near end of cell, followed by 2 silver spots toward apex; apex with black spot; silver spot at falcate indentation; 2-3 silver spots along tornus with basad spot of white with silver mesad; fringe fuscous, white distally with all white at falcate indentation; venter fuscous with dorsal white marks faintly repeated except distinct apical marks. **Hindwing:** fuscous; fringe fuscous; venter lustrous with some white scales on apical fringe. **Abdomen:** fuscous with silvery scales on posterior of each segment; venter mostly white: coremata absent. **Genitalia** (Fig. 173): tuba analis long; tegumen narrow, very thin dorsally; vinculum subrectangular, narrowing to distal end which is concave by saccus; saccus short, narrow; valva moderate, setaceous, very oblong with rounded but centrally truncate termen; short tubular anellus; aedeagus (Fig. 174) elongate (subequal to valval length), thin; cornutus a small tubule; vesica with spicules; phallobase short; ductus ejaculatorius with a campanulate hood near aedeagus (8 preparations examined).

Female (Fig. 91).- 4.8-5.0 mm. forewing length. Similar to male. **Genitalia** (Fig. 237): ovipositor short; papilla analis moderate, setaceous; apophyses short, thin, with posterior pair thinner and longer than anterior pair; ostium bursae (Fig. 236) a broad funnel with more than half sclerotized as basal cup; ductus bursae thin, sclerotized, relatively short; ductus seminalis emergent from bursa proximal to ductus bursae; corpus bursa large, elongate, with small stub-like accessory bursa anteriorly; bursa with spicules but no distinct signum (2 preparations examined).

Types.- Holotype male: Silverton, [San Juan Co.], Colorado, 16-23 Jul, [9200'(=2800m.)], (USNM). Paratypes (19 males, 8 females): **Arizona.-** Pima Co.: Santa Catalina Mts., 15 Jul 1938 (1 male), Bryant (LACM). **Colorado.-** Huerfano Co.: 6 mi. [=9.6 km.] S. Cuchara, Cucharas Pass, 20 Jul 1977 (1 male), 9800' [=2990 m.], on **Juncus,** J. B.

Heppner (USNM).   San Juan Co.: Silverton, 16-23 Jul (4
males), [9200' (=2800 m.)], (USNM).   San Miguel Co.: Trout
Lake, 17 Jul 1937 (7 males, 7 females), 9800' [=2990 m.], A.
B. Klots (AMNH).   New Mexico.- Rio Arriba Co.:   Lower
Canjilon Lakes, 7 mi. [=11.2 km.] NE. Canjilon, 29 Jul 1977
(5 males), 9800' [=2990 m.], J. P. & K. E. Donahue (LACM).
Taos Co.: Taos Ski Area, Sangre de Cristo Mts., 19 Jul 1977
(1 male), 9300' [=2840 m.], on Juncus, J. B. Heppner
(USNM).   Utah.- Wasatch Co.: 15 mi. [=24 km.] SE. Heber
City, 7 Jul 1976 (1 male, 1 female), on Juncus, J. B. Hep-
pner (JBH). (Paratypes to BMNH and CNC).

   Distribution (Fig. 11).- Colorado to New Mex.'co, Utah to
Arizona.

   Flight period.- July.

   Hosts.- Unknown (probably Juncus sp. (Juncaceae)).

   Biology.- This species has not been reared but adults
have been collected in proximity to creekside Juncus, mainly
over 9000 feet elevation.

   Remarks.- Superficially this species is difficult to
distinguish from G. juncivora and has long been mixed in
collections as G. californiae. Generally the midwing costal
margin white marks are distinctly thinner in G. roenastes
than in other species in the group. The genitalia, however,
are very distinct in both sexes. The vinculum of the male
varies somewhat in the development of the concave edge by
the saccus. The specimen illustrated (Fig. 173) has very
sharp lateral ends to the concave edge, while other specimens have the ends more rounded. The valvae also vary in
the degree of truncation distally.

   It is uncertain if G. roenastes will remain in this
species-group in the future since it appears to be somewhat
midway between the californiae and the montisella species-
group, having wing maculation similar to the former and
genitalia more like the latter, at least in the small accessory
bursa of the female. Since the accessory bursa is much
smaller than that found in species of the montisella species-
group and since the wing maculation is essentially the same
as in G. juncivora, the species is retained in the californiae
species-group for the present.

   The specific name is derived from Greek for "stream
dweller."

The montisella species-group

Forewing elongate–oblong, with falcate indentation following rounded acute apex. Hindwing pointed, without distinct termen. Labial palpus long, relatively smooth-scaled ventrally, with pointed apical segment slightly longer than 2nd segment, dorso–ventrally flattened. Eye moderate. Abdomen with coremata in 2 species (males) are unknown in 3 species. Valva oblong, generally elongate, setaceous, usually with saccular point; tegumen entire; tuba analis long with lateral sclerotized ridge each side; tubular anellus often strongly sclerotized; saccus usually short, narrow, without extension into vinculum; aedeagus with phallobase; ductus ejaculatorius with campanulate hood near aedeagus. Ovipositor moderate to short; papilla analis simple, setaceous; apophyses usually thin, short; ostium bursae a small cone-shaped sclerotized cup; ductus bursae sclerotized, short, thin, attached to bursa postero–ventrally; ductus seminalis emergent from bursa proximal to ductus bursae; corpus bursa usually elongate with accessory bursa; signum absent.

This species-group may be restricted to North America, including northern Mexico. Nine species are included in this group, 7 of which are new. There are several additional undescribed species in Mexico. The species generally have 6 white marks on the forewing costal margin, although some have only 5 and there are rare variations to 7 marks, but they all lack basal dorsal margin markings, while usually having a thin anal margin white line. The genitalia are much more diverse than the californiae species-group. The females differ from the previous species-group by possessing a well-developed accessory bursa.

The species-group appears most related to the californiae species-group. Inasmuch as 3 species are known only from females and one only from the male, the arrangement of the species in the group is probably not final; when these missing sexes are discovered, further assessment can be made of the relationships among the species.

Glyphipterix chiricahuae Heppner, new species
(Figs. 12, 102–103, 177–178, 240–241)

The distinct white line along the base of the dorsal margin of the forewings distinguishes the species from other North American Glyphipterix; the genitalia are also very distinct.

Male (Fig. 102).- 4.5-5.0 mm. forewing length. **Head:**
gray fuscous with lateral white eye margin; labial palpus
white dorsally, venter with basal segment white, 2nd segment with 2 alternating bands of black and white, and apical
segment black with white laterally and black and white
bands basally; antenna fuscous dorsally. **Thorax:** gray
fuscous; patagia gray fuscous; venter white with some fuscous; legs fuscous with white at joints. **Forewing:** gray
fuscous ground color with fuscous as borders to markings and
buff overlaid on apical half of wing; costal; margin with 6
white bars from basal crescent mark at 1/3 to apex in decreasing lengths, all with silver spots mesad; crescent mark
faintly extended to convergent 5th costal mark with similar
but mostly silver dorsal margin white mark at midwing;
black spot at end of cell, followed distad by 2 silver spots;
apex with black spot; falcate indentation with silver spot;
termen with 3 (often merged) silver spots; tornus with small
white spot having silver bar mesad; fringe fuscous, white
distally except all white at falcate indentation; venter
fuscous with dorsal white marks faintly repeated except
distinct apical marks. **Hindwing:** fuscous; fringe fuscous;
venter fuscous with white scales at apex. **Abdomen:** fuscous
with silver scales on posterior each segment; venter same
but broad white scale row posterior of each segment; coremata absent. **Genitalia** (Fig. 177): tuba analis long;
tegumen stout; vinculum subrectangular with concave saccus
edge; saccus short, narrow; valva elongate–oblong, setaceous,
with rounded termen recurved to long saccular point somewhat ventrally directed; aedeagus (Fig. 178) elongate (subequal to valval length), thin; cornutus a short tubule (2
preparations examined).

Female (Fig. 103).- 4.6-4.9 mm. forewing length. Similar
to male. **Genitalia** (Fig. 241): ovipositor short; papilla analis
elongate, setaceous; apophyses short, with posterior pair
somewhat longer than anterior pair and anterior pair stouter;
ostium bursae (Fig. 240) a deep sclerotized cup; ductus bursae thin, sclerotized, somewhat long; corpus bursae large,
elongate–ovate, with moderate accessory bursa (not readily
discernible in figure) (2 preparations examined).

Types.- Holotype male: S. W. R. S. [=Southwestern Res.
Sta.], 5 mi.[=8 km.] W. Portal, [Chiricahua Mts.], Cochise
Co., Arizona, 9 Nov 1964, 5400' [=1650 m.], V. D. Roth
(SWRS). Paratypes (3 males, 4 females): **Arizona.**- same
data as holotype (2 males, 4 females) (SWRS), 7 Oct 1964 (1

male), V. D. Roth (SWRS). (Holotype to AMNH; paratypes to AMNH, BMNH, and USNM).

Distribution (Fig. 12).- Known only from southeast Arizona.

Flight period.- October to November.

Hosts.- Unknown.

Biology.- Unknown.

Remarks.- The forewing dorsal margin white mark at midwing is less extended than in other species of the montisella species-group but has a longer silver extension. The very distinct dorsal margin white line is characteristic of the species. The ostium is deeper and narrower than in the other species of the group.

Glyphipterix chiricahuae is the only species of the group with adults active in autumn, as far as is known. Since the adults fly so late in the year, the species may be more widespread but not collected due to lesser collecting activities in the autumn season. The species is closely related to the new species from Flagstaff, Arizona, Glyphipterix hodgesi, new species, but it is not known how closely the relationships are to the other species of the group.

The species is named after the Chiricahua Mountains of southeastern Arizona.

Glyphipterix hodgesi Heppner, new species
(Figs. 12, 108-109, 179-180, 242-243)

The male genitalia readily distinguish the species while the forewing maculation is similar to the previous species.

Male (Fig. 108).- 4.3-4.9 mm. forewing length. Head: fuscous with white lateral eye margin; labial palpus white dorsally, venter with basal segment white, 2nd segment with 2 alternating bands of black and white, and apical segment black and white laterally and basal band of black and white; antenna fuscous dorsally. Thorax: fuscous; patagia fuscous; venter white with some fuscous; legs fuscous with white at joints. Forewing: fuscous ground color with gray fuscous on basal 1/3 and buff overlaid on apical half except for fuscous borders to markings; dorsal margin with reduced thin white line from 1/4 to base; costal margin with 6 white bars from basal 1/3 to apex in decreasing lengths each with silver spot mesad except basal crescent-shaped mark which points to convergence of next white mark, with equal white and silver mark from dorsal margin; fuscous spot at end of cell, followed distad by 2 silver spots; black spot at apex; silver spot at falcate indentation; 2-3 silver spots (sometimes merged) along termen; tornus with white spot with silver mesad; fringe fuscous, white distally except all white at falcate indentation; venter fuscous with dorsal white marks faintly repeated except distinct apical marks. **Hindwing:** fuscous; fringe fuscous; venter whitish fuscous with white scales on apical fringe. **Abdomen:** fuscous with silvery scales on posterior of each segment; venter same but mostly white on posterior of each segment; coremata absent. **Genitalia** (Fig. 179): tuba analis long; tegumen stout; vinculum subrectangular with concave edge by saccus; saccus short, narrow; valva elongate-oblong, setaceous, with termen rounded to ventrad projecting saccular extension; aedeagus (Fig. 180) elongate (subequal to valval length), thin; cornutus a short tubule (3 preparations examined).

Female (Fig. 109).- 4.5-4.9 mm. forewing length. Similar to male. **Genitalia** (Fig. 243): ovipositor short; papilla analis elongate, setaceous; apophyses short, thin, with posterior pair twice length of anterior pair, anterior to basal fork; ostium bursae (Fig. 242) a wide cone-shaped sclerotized cup; ductus bursae moderate length, thin, sclerotized; corpus bursae elongate-ovate, large, with moderate accessory bursa (2 preparations examined).

Types.- Holotype male: Hart Prairie, [San Francisco Mts.], 10 mi. [=16 km.] NW. Flagstaff, Coconino Co., Arizona, 23 Aug 1961, 8500' [=2590 m.], R. W. Hodges (USNM). Paratypes (9 males, 3 females); **Arizona.-** Coconino Co.: same locality as holotype, 18 Aug 1961 (1 male), 22 Aug 1961 (1 male), 25 Aug 1961 (1 male), 28 Aug 1961 (1 female), 30 Aug 1961 (3 males), R. W. Hodges (USNM); Fort Valley, 7.5 mi. [=12 km.] NW Flagstaff, 21 Aug 1961 (1 male), 23 Aug 1961 (1 male), 28 Aug 1961 (1 male), 7350' [=2240 m.], R. W. Hodges (USNM); West Fork, 16 mi. [=25.6 km.] SW. Flagstaff, 10 August 1961 (2 females), 6500" [=1980 m.], R. W. Hodges (USNM). (Paratypes to BMNH, CNC, UCB).

Distribution (Fig. 12).- Known only from the San Francisco Mountains and the Flagstaff plateau of north central Arizona.

Flight period.- August.

Hosts.- Unknown.

Biology.- Unknown.

Remarks.- Superficially **G. hodgesi** is very similar to **G. chiricahuae** but the dorsal margin white line is more re-

duced. As in G. **chiricahuae**, the species also has the midwing dorsal white bar short with a long silver extension.
The genitalia of the two species are very distinct, yet show
close relationships.

The species is named in honor of Dr. Ronald W. Hodges,
Systematic Entomology Laboratory, USDA, and collector of
all the specimens of this species.

Glyphipterix saurodonta Meyrick
(Figs. 13, 104, 248-249)

Glyphipteryx [sic] **saurodonta** Meyrick, 1913a:68;
    1913b:44; 1914c:30; Barnes & McDunnough, 1917:
    182; Forbes, 1923:355; McDunnough, 1939:84;
    Braun, 1940:274; Procter, 1946:318; Clarke,
    1955:277 (index).
Glyphipterix saurodonta.- Clarke, 1969:80; Heppner,
    1982a:52; 1983a:26.

This species is distinct in lacking a long mesal point on
the basal costal crescent mark of the forewings and in lacking a very distinct white line on the anal margin.

Male.- Unknown.
Female (Fig. 104).- 3.8-4.3 mm. forewing length. **Head:**
fuscous with white line at eye margin; labial palpus dorsally
white, venter with basal segment white, 2nd segment with 2
alternating bands of fuscous and white, and apical segment
fuscous with white laterally and fuscous and white band at
base; antenna fuscous dorsally. **Thorax:** fuscous; patagia
fuscous; venter white and fuscous; legs fuscous with white at
joints. **Forewing:** ground color fuscous with apical half
overlaid with buff scales except for fuscous borders of
markings; some gray overlaid scales on base of wing; very
narrow white line along anal margin; costal margin with 6
white bars from 1/3 of wing to apex in decreasing lengths,
each except basal mark with a silver spot mesad; 5th white
mark from apex nearly meeting similar white and silver
mark of dorsal margin; fuscous spot near end of cell; 2 silver
spots beyond cell; apex with black spot; silver spot at falcate indentation; 2 silver spots on tornus, with white and
silver spot on basal end of tornus; fringe fuscous, white
distally except all white at falcate indentation; venter
fuscous with dorsal white marks faintly repeated except
distinct apical marks. **Hindwing:** fuscous; fringe fuscous;

venter lustrous pale fuscous with a few white scales on
apical fringe. **Abdomen:** fuscous with silvery scales on
posterior of each segment; venter same but mostly white.
Genitalia (Fig. 249): ovipositor short; papilla analis
moderate, setaceous; apophyses short, thin, posterior pair
somewhat longer than anterior pair; ostium bursae (Fig.
248) an elongate sclerotized cup with a distinct concave posterior
margin on 7th sternite; ductus bursae short, thin, sclerotized;
ductus seminalis emergent from bursa proximal to ductus
bursae; corpus bursae elongate, large with large accessory
bursa anteriorly; signum absent (3 preparations examined).

　　Type.- Holotype female: Toronto, Ontario, Canada, Sep
1912, Parish (BMNH).

　　Additional specimens (4 females).- **New York.-** St. Law-
rence Co.: Potsdam, 1897 (1 female), (MCZ). Tompkins Co.:
McLean, 21 Sep 1930 (1 female), (AMNH). **West Virginia.-**
Hardy Co.: Lost River St. Park, 19 Sep 1938 (1 female), A.
F. Braun (ANSP). **Canada.- Ontario.-** Ottawa, 17 Aug 1906
(1 female), C. H. Young (CNC).

　　Distribution (Fig. 13).- Collection records are for southern Ontario, New York, and West Virginia.

　　Flight period.- Mid-August to September.

　　Hosts.- Unknown.

　　Biology.- Unknown.

　　Remarks.- This species is rare in collections, possibly
due to its late summer adult emergence. It superficially
resembles a species from Arizona, **G. chiricahuae,** but is
smaller and lacks the more distinct anal margin white line
present in the Arizona species. The genitalia of the two
species are quite distinct. When the male is discovered it
will be possible to more accurately determine which species
actually are most closely related to **G. saurodonta.**

Glyphipterix cherokee Heppner, new species
(Figs. 13, 105, 254–255)

　　This eastern species is distinguished primarily by the
short cup-shaped ostium bursae, while wing maculation resembles a dark **Glyphipterix montisella** Chambers.

　　Male.- Unknown.
　　Female (Fig. 105).- 5.3 mm. forewing length. Head:
fuscous with lateral white eye margin line; labial palpus
white dorsally, venter with basal segment white, 2nd segment with 2 alternating bands of fuscous and white, and

apical segment fuscous with white laterally; antenna dorsally fuscous. **Thorax:** fuscous; patagia fuscous; venter fuscous and white; legs fuscous with white at joints. **Forewing:** fuscous ground color with buff and copper-iridescent scales overlaid on apical half of wing except for fuscous borders to markings; anal margin with narrow white line to base; costal margin with 6 white bars from basal 1/3 of wing to apex in decreasing lengths, with basal mark a curved crescent pointed to convergence of midwing silver spots of dorsal and costal margin white marks, silver spots mesad of all other white bars; dark fuscous at end of cell, then 2 silver spots thereafter; tornus with 2 silver spots; then a white and silver spot on basal end of tornus; silver spot at falcate indentation; apex with a black spot; fringe fuscous, white distally except all white at falcate indentation; venter fuscous with faintly repeated dorsal white marks except for distinct apical marks. **Hindwing:** fuscous; fringe fuscous; venter pale fuscous with white on apical fringe. **Abdomen:** fuscous with silvery scales on posterior of each segment; venter same but mostly white. **Genitalia (Fig. 255):** ovipositor short; papilla analis moderate, setaceous; apophyses short, thin, posterior pair near twice length anterior pair; ostium bursae (Fig. 254) a small wide, sclerotized cup; ductus bursae short, thin, sclerotized; ductus seminalis emergent from bursa proximal to ductus bursae; corpus bursae relatively large, elongate, with anterior moderate accessory bursa; signum absent (1 preparation examined).

Type.- Holotype female: Great Smoky Mts., Tennessee, 23 Aug 1950, 6000' [=1830 m.], G. S. Walley (CNC).

Distribution (Fig. 13).- Tennessee; probably also North Carolina.

Flight period.- August.

Hosts.- Unknown.

Biology.- Unknown.

Remarks.- This species is known only from a unique female from the southern Appalachians. The genitalia are rather distinctive in the shape of the ostium, which resembles only the ostium of **Glyphipterix chambersi**, but without the male no clear relationship is evident. The wing maculation resembles **G. montisella**.

The specific name is named after the Cherokee Indians of the type locality.

Glyphipterix chambersi Heppner, new species
(Figs. 13, 106, 256–257)

A small eastern species related to **Glyphipterix monti-sella** but smaller and typically with only 5 white marks on the forewing costal margin.

Male.- Unknown.
Female (Fig. 106).– 4.8–5.5 mm. forewing length. **Head:** fuscous, with white eye margin reduced; labial palpus dorsally white, venter with basal segment white, 2nd segment with 2 alternating bands of fuscous and white, and apical segment fuscous with white laterally.; antenna dorsally fuscous. **Thorax:** fuscous; patagia fuscous; venter silvery fuscous; legs fuscous with white at joints. **Forewing:** fuscous ground color with buff overlaid on apical half except fuscous borders of markings; anal margin with narrow white line; costal margin with 5 (rarely 6) white bars from basal 1/3 of wing to apex in decreasing lengths, with 1st mark a crescent curved to convergent (or nearly convergent) next distal costal bar having equal bar on dorsal margin, silver spot between, and other white bars with silver spot mesad; 2 silver spots beyond cell, with dark fuscous at end of cell; black spot on apex; silver spot at falcate indentation; 2 silver spots along tornus with 3rd silver and white spot basad of tornus; fringe fuscous, white distally and all white at falcate indentation; venter fuscous with dorsal white marks faintly repeated except distinct apical marks. **Hindwing:** fuscous; fringe fuscous; venter fuscous with some white on apical fringe. **Abdomen:** fuscous with silvery scales on posterior of each segment; venter mostly white. **Genitalia** (Fig. 257): ovipositor short; papilla analis elongate, setaceous; apophyses short, thin; with posterior pair nearly twice length anterior pair, anterior to basal fork; ostium bursae (Fig. 256) a wide sclerotized cup; ductus bursae thin, short, sclerotized; ductus seminalis emergent from bursa proximal to ductus bursae; corpus bursae relatively large, elongate with large accessory bursa anteriorly; signum absent (3 preparations examined).

Types.- Holotype female: R'thwaite [=Roundthwaite], Marmont, Manitoba, Canada, 8 Aug 1905 (USNM). Paratypes (6 females, 4 others without abdomens): **Kentucky.-** [Kenton Co.: Covington?], (4 females, 3 others without abdomens, MCZ; 1 female, USNM; 1 without abdomen, LACM), V. T. Chambers. **Canada.- Manitoba.-** R[oun]thwaite, Marmont, 8 Aug 1905 (1 female) (USNM). (There are 4 other specimens from Kentucky in the MCZ collection with only remnants, wing or thorax, and these are not counted above nor selected as paratypes).

Distribution (Fig. 13).- Known only from Manitoba and Kentucky.

Flight period.- August.

Host.- Unknown.

Biology.- Unknown.

Remarks.- This species is very close to Glyphipterix montisella but usually has only 5 white marks on the forewing costal margin, while G. montisella usually has 6 white marks (although rarely also 5) and is larger in wingspread on average. In the female the ostium bursae is wider in G. chambersi than the more narrow and anteriorly more pointed ostium of G. montisella. The two specimens of G. chambersi that have 6 white forewing costal marks resemble G. saurodonta but do not have a posterior distinct edge to the 7th sternite. One of the 6-marked G. chambersi specimens from Kentucky has only 5 marks on the right forewing, thus indicating that this variation is unusual in the species. The holotype has been selected from Manitoba primarily because the Kentucky specimens are in poor condition.

The species is named in honor of Dr. Vactor T. Chambers, the microlepidopterist from Kentucky who first collected this species over 100 years ago.

Glyphipterix montisella Chambers
(Figs. 14, 110–111, 181–182, 244–245)

Glyphipteryx [sic] montisella Chambers, 1875b:292; 1877b:129; 1877c:143; 1877d:149; 1878a:116; 1878b:148: 1878c:148; Riley, 1891:104; Dyar, 1900:84; [1903]:493; Kearfott, 1903:108; Barnes & McDunnough, 1917:182; McDunnough, 1939:84.

Glyphipteryx [sic] montinella [sic] Chambers, 1877a: 14, missp.

Glyphipteryx [sic] montella Meyrick, 1913b:44, emend.; 1914c:30.

Glyphipterix montisella.- Heppner, 1982a:51; 1983a: 26.

A relatively large Rocky Mountain and Great Basin species typically with 6 forewing costal margin white marks (rarely 5 or 7) and with male valvae with a rounded indentation by the saccular point.

Male (Fig. 110).- 5.6-6.4 mm. forewing length. **Head:** fuscous with white line along lateral eye margin; labial palpus dorsally white with some fuscous merging laterally from ventral markings, with venter basal segment white, 2nd segment with 2 alternating bands of black or dark fuscous and white, and apical segment with basal black and white bands, followed by black and white laterally; antenna dorsally fuscous. **Thorax:** fuscous; patagia fuscous; venter shining fuscous and white; legs fuscous with white at joints. **Forewing:** fuscous ground color with yellow-buff overlaid on apical half except for borders of markings; gray fuscous area on basal 1/4 from CuP fold to dorsal margin; anal margin white line usually absent or vestigial; costal margin with 6 (rarely 5 or 7) white bars from 1/3 of wing to apex with basal mark a long crescent curved to near convergence of next distal bar with same mark from dorsal margin, both with silver spots mesad; silver spots mesad of white marks except basal crescent; fuscous spot at end of cell, followed by 2 silver spots; apex with black spot; silver spot at falcate indentation; silver bar and spot along tornus to basal white spot with silver mesad; fringe fuscous, white distally except all white at falcate indentation; venter fuscous with dorsal white marks faintly repeated except distinct apical marks. **Hindwing:** lustrous fuscous; fringe fuscous; venter fuscous with white at apical fringe. **Abdomen:** fuscous with silvery scales on posterior of each segment; venter same but white scale row posterior of each segment; coremata absent. **Genitalia (Fig. 181):** tuba analis long, with strong lateral sclerotized ridges; tegumen narrow; vinculum subrectangular with concave edge at saccus; saccus short narrow; valva elongate-oblong with rounded invagination to sclerotized saccular tip, setaceous; strongly sclerotized tubular anellus; aedeagus (Fig. 182) elongate (subequal to valval length), thin; cornutus a short tubule; vesica with spicules; phallobase small (10 preparations examined).

Female (Fig. 111).- 4.9-5.2 mm. forewing length. Similar to male. **Genitalia (Fig. 245):** short ovipositor; papilla analis elongate, setaceous; apophyses short, somewhat stout, with posterior pair near twice length anterior pair, anterior to basal fork; ostium bursae (Fig. 244) a deep sclerotized cup; ductus bursae short, thin, sclerotized; corpus bursae large; elongate with moderate accessory bursa (6 preparations examined).

Type.- Lectotype male, by present designation: [Spanish Bar (=Fall R.)], Denver, South Park, [no date], 10,000'

[=3050 m.], [V.T. Chambers] (MCZ). Chambers (1875b) noted that his description was based on many specimens he had collected in Colorado at Spanish Bar. Only one specimen, taken to be a syntype, has been located of these Colorado specimens (identified by characteristic pin and labels typical of Chambers specimens) and is selected as lectotype and labelled as such).

Additional specimens (74 males, 17 females).- **Arizona.**- Coconino Co.: Flagstaff, 19 Jul 1939 (1 male), A. F. Braun (ANSP); Hart Prairie, 10 mi [=16 km.] NW. Flagstaff, [San Francisco Mts.], 30 Aug 1961 (1 male), 8500' [=2590 m.], R. W. Hodges (USNM). **California.**- Mono Co.: Rock Cr., 1 mi. [=1.6 km.] W. Toms Place, 9 Aug 1959 (2 males), C. D. MacNeill (CAS). **Colorado.**- Denver, [no date] (1 male), Oslar (LACM); South Park, [no date] (1 female), Oslar (LACM). Boulder Co.: Univ. of Colo., [Boulder], Aug (1 male), Cockerell (USNM). Larimer Co.: Hidden Valley, Rocky Mtn. Natl. Park, 11 Aug 1929 (10 males, 1 female, ANSP; 7 males, 3 females, USNM), 12 Aug 1929 (1 male, 1 female, ANSP; 1 male, USNM), A. F. Braun. Routt Co.: Dry Lake Cogd., 5 mi. [= 8 km.] NE. Steamboat Springs, 26 Jul 1977 (1 male), 8000' [=2440 m.], E. Randal & J. A. Powell (UCB). San Juan Co.: Silverton, 8-15 Jul (1 male), 16-23 Jul (1 male), 1-7 Aug (1 male), 9200' [=2800 m.], (USNM). **Montana.**- Glacier Natl. Park, Rising Sun, 2 Aug 1973 (2 males), 4500' [=1370 m.], E. Jäckh (USNM). **New Mexico.**- San Miguel Co.: Cowles, Sangre de Cristo Mts., 22 Jul 1939 (5 males, 1 female, ANSP), 23 Jul 1939 (2 males, ANSP; 1 female, USNM), 24 Jul 1939 (2 males, 2 females, ANSP; 1 female, USNM), 27 Jul 1939 (1 female, ANSP), A. F. Braun. **Utah.**- Cache Co.: Logan, Aug 1933 (1 male) (AMNH). Utah Co.: Timpooneke Cpgd., Mt. Timpanogos, 7 Jul 1976 (7 females), J. B. Heppner (JBH); 29 Jul 1967 (9 males, 3 females), 7400' [=2260 m.], J. A. Powell (UCB). Wasatch Co.: 15 mi [=24 km.] SE. Heber City, 7 Jul 1976 (1 male), J. B. Heppner (JBH). **Wyoming.**- Yellowstone Natl. Park, Old Faithful, 5 Aug 1934 (1 male), A. F. Braun (ANSP). Albany Co.: 1.5 mi. [=2.4 km.] E. Centennial, 25 Jul 1977 (16 males), 8050' [=2450 m.], on *Juncus*, J. B. Heppner (USNM).

Distribution (Fig. 14).- Collection records are from Montana to New Mexico, Utah to Arizona, and the eastern slope of the Sierra Nevada in California, generally at high elevations.

Flight period.- July to August.

Hosts.- ?**Juncus** sp. (Juncaceae).

Biology.- The species has not been reared but is usually collected in close proximity to Juncus, the presumed host plant, in marshy areas or along creeks.

Remarks.- Glyphipterix montisella varies in forewing costal white marking but usually has 6 marks, while rarely having an additional small white spot as a new 3rd mark from the apex or a reduction (usually in females) to 5 marks with the normal 4th from the apex being vestigial or absent. Some specimens have a small amount of buff on the anal quarter of the forewing near the cubital fold but this is never as large and distinct as in Glyphipterix flavimaculata, new species, described next.

The species is widespread in the Rockies, where it is usually encountered at high elevations. The specimens from California are from an area of the state essentially Great Basin in faunal composition.

Glyphipterix montisella is most closely related to G. flavimaculata and shows further close relationship to two new species from Arizona described subsequently. Glyphipterix chambersi appears to be close to G. montisella as well, especially those specimens of G. montisella with reduced forewing costal marks.

Glyphipterix flavimaculata Heppner, new species
(Figs. 14, 107, 162-163)

This southern California species is readily distinguished by the large basal buff mark and the reduction of yellow scaling on the apical quarter of the forewing.

Male (Fig. 107).- 6.1 mm. forewing length.   Head: fuscous with lateral white eye margin; labial palpus white dorsally, venter with basal segment white, 2nd segment with 2 alternating bands of fuscous and white, and apical segment with basal fuscous and white bands, then fuscous with white laterally; antenna fuscous dorsally. Thorax: fuscous; patagia fuscous; venter white with some fuscous; legs fuscous with white at joints. Forewing: fuscous ground color overlaid with yellow-buff midwing between white marks and their fuscous borders; apical 1/3 of wing fuscous; basal 1/3 with large ovate buff mark along CuP fold; thin buff line along anal margin; costal margin with 6 white marks, with basal mark a long crescent curved to near convergence of silver spots mesad of costal and dorsal white bars, all other white marks with silver spot mesad; 2 silver spots beyond end of cell; apex with a black spot; falcate indentation with silver spot; silver line and spot along tornus, then white and silver spot at base of tornus; fringe fuscous, white dorsally, with all white at falcate indentation; venter fuscous with dorsal white marks faintly repeated except distinct apical marks. **Hindwing:** fuscous; fringe fuscous; venter fuscous with some white on apical fringe. **Abdomen:** fuscous with silvery scales on posterior each segment; venter similar but white on posterior of each segment; rudimentary abdominal coremata. **Genitalia** (Fig. 162): tuba analis long, narrow; vinculum subrectangular with concave edge by saccus; saccus short, narrow; valva elongate–oblong, setaceous, with slight invagination by saccular point; strongly sclerotized tubular anellus; aedeagus (Fig. 163) short (less than valval length), thin; cornutus a short tubule; vesica with spicules; phallobase small (1 preparation examined).

Female.- Unknown.

Type.- Holotype male: Hathaway Cr., San Bernardino Mts., [San Bernardino Co.], California, 26 Jul 1942, 8000' [=2440 m.] (USNM).

Distribution (Fig. 14).- Known only from southern California.

Flight period.- July.

Hosts.- Unknown.

Biology.- Unknown.

Remarks.- This species is very closely related to **Glyphipterix montisella** but differs strikingly in the large basal buff mark of the forewings, as well as the reduction of the yellow–buff on the apical half of the forewings. The male genitalia are similar but differ in the very small invagination near the saccular point and by the shorter aedeagus.

The specific name is derived from Latin for the buff basal spot of the forewings.

Glyphipterix melanoscirta Heppner, new species
(Figs. 15, 116, 164)

A small, nearly black species from Arizona with 6 forewing costal marks and a straight basal mark on the costal margin.

Male (Fig. 116).- 3.6 mm. forewing length. **Head:** fuscous with white lateral eye margin; labial palpus white dorsally, venter with basal segment white, 2nd segment with 2 alternating bands of fuscous and white, and apical segment fuscous with white laterally and basal band of fuscous and white; antennal dorsally fuscous. **Thorax:** fuscous; patagia fuscous; venter white with fuscous; legs fuscous with white at joints. **Forewing:** dark fuscous ground color; costal margin with 6 white bars from basal 1/3 to apex in decreasing lengths, with basal mark straight but oblique; costal white mark 5th from apex oblique, with silver mesad spot merging with similar silver and white bar from dorsal margin; silver spots mesad of each of apical 5 white costal margin marks except basal mark; black spot at end of cell, followed by 2 silver spots distad; apex with black spot; silver spot at falcate indentation; termen with silver line and spot, then a white spot on tornus with silver mesad; fringe fuscous, white distally except all white at falcate indentation; venter gray fuscous with dorsal white marks faintly repeated except for distinct apical marks. **Hindwing:** fuscous; fringe fuscous; venter fuscous with white scales on apex.   **Abdomen:** dark fuscous with silvery scales on posterior of each segment; venter same but white scales on posterior of each segment; abdominal coremata present.   **Genitalia** (Fig. 164): tuba analis long; tegumen narrow; vinculum subrectangular with concave edge at saccus; saccus moderate, narrow; valva elongate–oblong, setaceous, with deep invagination by long saccular point; tubular anellus strongly sclerotized; aedeagus (Fig. 164, in situ), elongate (nearly equal to valval length), thin; cornutus a short tubule; vesica with spicules; phallobase small (1 preparation examined).

   Female.- Unknown.

   Type.- Holotype male: S. W. R. S. [=Southwestern Res. Sta.], 5 mi. [=8 km.] W. Portal, [Chiricahua Mts.], Cochise Co., Arizona, 21 Jul 1967, 5400' [=1650 m.] (AMNH).

   Distribution (Fig. 15).- Known only from southeastern Arizona.

   Flight period.- July

   Hosts.- Unknown.

   Biology.- Unknown.

   Remarks.- The small size and lack of buff scales on the forewing, being overall dark fuscous, readily distinguishes this species from **Glyphipterix santaritae**, new species, and the genitalia differ from other North American species of the genus.   The basal costal margin white mark is also straight and not a crescent as in the other related species.

   The specific name is derived from Greek for "black leaper" and refers to the behavior observed for other glyphipterigids.

Glyphipterix santaritae Heppner, new species
(Figs. 15, 112-113, 183-184, 246-247)

An Arizona species with a distinct crescent mark on the basal costal margin of the forewings, while genitalia are similar to G. melanoscirta.

Male (Fig. 112).- 4.2-5.0 mm. forewing length. **Head:** fuscous with white lateral eye margin; labial palpus white dorsally, venter with basal segment white, 2nd segment with 2 alternating bands of black and white, and apical segment black with white laterally and a basad band of black and white; antenna fuscous dorsally. **Thorax:** fuscous; patagia fuscous; venter white with fuscous; legs fuscous with white at joints. **Forewing:** fuscous ground color with yellow-buff scales overlaid on apical half except for fuscous border of markings; costal margin with 6 white bars from basal 1/3 to apex in decreasing lengths, with basal mark a crescent curved to near convergence of next white bar with similar mark on dorsal margin; white marks except crescent with silver mesad spot each; fuscous spot at end of cell, followed distad by 2 silver spots; black spot on apex; silver spot at falcate indentation; termen with silver bar and spot; tornus with a small white mark with silver mesad; fringe fuscous, white distally with all white at falcate indentation; venter fuscous with dorsal white marks faintly repeated except for distinct apical marks. **Hindwing:** fuscous; fringe fuscous; venter fuscous with a few white scales on apical fringe. **Abdomen:** fuscous with silvery scales on posterior of each segment; venter same but white scales on posterior of each segment; coremata present. **Genitalia (Fig. 183):** tuba analis long; tegumen stout; vinculum subrectangular with concave saccus edge; saccus moderate, narrow; valva elongate-oblong, setaceous, with deep invagination by long saccular point; strongly sclerotized tubular anellus; aedeagus (Fig. 184) elongate (subequal to valval length), thin; cornutus a short tubule; vesica with spicules (2 preparations examined).

Female (Fig. 113).- 4.5-4.8 mm. forewing length. Similar to male. **Genitalia (Fig. 247):** ovipositor short; papilla analis long, setaceous; apophyses short, moderately stout, posterior pair somewhat less than twice length of anterior pair anterior to basal fork; ostium bursae (Fig. 246) a sclerotized cup; ductus bursae short, thin, sclerotized; corpus bursae relatively large, elongate with moderate accessory bursa (2 preparations examined).

Types.- Holotype male: Madera Cyn., Santa Rita Mts., Santa Cruz Co., Arizona, 1-3 Aug 1970, P. Rude (UCB). Paratypes (4 males, 2 females): Arizona.- same data as holotype (1 female) (UCB). Cochise Co.: E. Turkey Cr., Chiricahua Mts., 10 Aug 1972 (3 males), on Potentilla flowers, J. T. Doyen (UCB). Pima Co.: Bear Cyn., Santa Catalina Mts., 2 Aug 1970 (1 female), J. A. Powell (UCB); Santa Catalina Mts., 15 Jul 1938 (1 male), Bryant (LACM). (Holotype to CAS; paratype to USNM).

Distribution (Fig. 15).- Known only from southeastern Arizona.

Flight period.- August.

Hosts.- Unknown.

Biology.- Unknown. The species has been collected on flowers of Potentilla sp. (Rosaceae).

Remarks.- Of the two Glyphipterix having a deeper valval invagination than in the related G. montisella, G. santaritae has the deepest invagination and has a distinct crescent mark on the costal margin, versus the straight basal mark of G. melanoscirta. One specimen collected in the Santa Catalina Mountains in 1938 has only 5 costal marks but is otherwise the same.

The specific name is derived from the name of the Santa Rita Mountains of the type locality.

The ruidosensis species-group

Forewing elongate-oblong with falcate indentation following rounded but acute apex. Hindwing pointed without distinct termen. Labial palpus long, relatively smooth-scaled ventrally, with pointed apical segment somewhat longer than 2nd segment, dorso-ventrally flattened. Eye moderate. Abdomen without coremata. Valva highly modified, strongly sclerotized, short, with long saccular point and strong setae on dorsal apex; tegumen entire but very thin dorsally; tuba analis long with lateral sclerotized ridge each side; tubular anellus sclerotized; saccus relatively short, narrow, without extension into vinculum; vinculum subrectangular with concave edge by saccus; aedeagus with phallobase; vesica with spicules; ductus ejaculatorius with campanulate hood near aedeagus. Ovipositor short, modified with sclerotized projections from 8th sternite; papilla analis simple, setaceous; apophyses short, stout; ostium bursae a cone-shaped sclerotized cup; ductus bursae thin, short, sclerotized, attached to bursa postero-ventrally; ductus seminalis emergent from bursa proximal to ductus bursa; bulla seminalis large; corpus bursae large, elongate–ovate, with accessory bursa; signum absent.

This species-group presently contains only one unusual species with highly modified genitalia. The forewing maculation has a dorsal margin basal mark similar to that found in the **californiae** species-group. The female genitalia show relationship to the **montisella** species-group in the type of bursa copulatrix and short ovipositor, while the male genitalia could be modified from either species-group, although more probably from species like G. **montisella**. Due to the unusual genitalia the group is placed after the **montisella** species-group.

Glyphipterix ruidosensis Heppner, new species
(Figs. 15, 114–115, 189, 258-259)

A New Mexican species with distinctive genitalia and with wing maculation similar to **Glyphipterix chiricahuae** but with a basal crescent on the forewing dorsal margin.

Male (Fig. 114).– 6.3 mm. forewing length. **Head:** fuscous with lateral white eye margin; labial palpus white dorsally, venter with basal segment white, 2nd segment with 2 alternating bands of black and white, and apical segment black with white laterally and a black and white band basally; antenna dorsally fuscous. **Thorax:** fuscous; patagia fuscous; venter white with some fuscous; legs fuscous with white at joints. **Forewing:** fuscous ground color with gray overlaid on basal half and yellow-buff on apical half except for fuscous borders to markings; dorsal margin basal 1/4 with long white crescent directed apically to near CuP fold; costal margin with 6 white bars from basal 1/3 to apex in decreasing lengths, each with a silver spot mesad and with basal mark a long crescent pointed to convergence of 5th from apex mark with similar mark from dorsal margin mid-wing; black spot at end of cell and on apex; 2 silver spots beyond end of cell and silver spot at falcate indentation; apical 1/4 with coppery shine; silvery line and spot along termen; tornus with white spot having silver bar mesad; fringe fuscous, white distally except all white at falcate indentation; venter fuscous with dorsal white marks faintly repeated except distinct apical marks. **Hindwing:** fuscous; fringe fuscous; venter whitish fuscous with white scales on apical fringe. **Abdomen:** fuscous with silver scales on pos-

terior of each segment; venter same but white on posterior
of each segment. **Genitalia** (Fig. 189): tegumen stout; vinculum with broad concave edge at saccus; saccus moderate,
narrow; valva short, strongly sclerotized, with dense field of
spine-like setae on valval dorsal tip, with large elongated
saccular point; aedeagus (Fig. 189, in situ) long (twice valval
length), thin, tapering to thinner apical area; cornutus a
short tubule (1 preparation examined).

    Female (Fig. 115).- 5.5 mm. forewing length. Similar to
male. **Genitalia** (Fig. 259): ovipositor short, with 8th
sternite modified by stubby sclerotized points from base of
anterior apophyses; papilla analis elongate, setaceous;
apophyses stout, posterior pair near twice length anterior
pair, anterior to basal fork; ostium bursae (Fig. 258) a deep,
sclerotized cone-like cup; ductus bursae sclerotized, thin,
short; corpus bursae elongate-ovate, large with moderate
accessory bursa (1 preparation examined).

    Types.- Holotype male: Ruidoso Canyon, [Sacramento
Mts., Otero-Lincoln Co.], New Mexico, 20 Sep 1916, C.
Heinrich (USNM). Paratype (1 female): same locality as
holotype, 4 Oct 1916 (1 female), C. Heinrich (USNM).

    Distribution (Fig. 15).- Known only from southern New
Mexico.

    Flight period.- Late September to October.

    Hosts.- Unknown.

    Biology.- Unknown.

    Remarks.- The modified genitalia of this species allow
easy identification, while the forewing maculation is distinct
from the **montisella** species-group by the addition of the
basal white mark of the dorsal margin. The 6 white marks
of the costal margin also distinguish G. **ruidosensis** from the
somewhat similar appearing G. **feniseca** of California.

    There is no other known species that has genitalia even
somewhat similar to G. **ruidosensis** and a separate species-
group has been erected for it.

    The type locality involves a long canyon with upper
reaches on the slopes of Sierra Blanca Peak (Otero Co.),
rising to over 12,000 feet in elevation, while lower portions
(Lincoln Co.) pass through montane coniferous forest before
reaching the town of Ruidoso at about 8,000 feet elevation
and then continue toward the eastern New Mexican plains at
6,000 feet elevation and less. The elevation of the actual
collecting site and the corresponding habitat along the
canyon are not known, although very likely near Ruidoso.

    The specific name is based on the name of the type lo-
cality, Ruidoso Canyon.

Diploschizia Heppner, 1981
(Figs. 28-29, 35, 41, 47, 57, 130-131, 280-288, 294-298)

Diploschizia Heppner, 1981a:309 (type-species: Glyphipteryx [sic]   impigritella Clemens, 1863,
original designation); Heppner, 1982a:54, 1982b:
249: 1984:55.

The lack of M3 in the hindwings distinguishes this genus
from otherwise similar Glyphipterix species.

Adults small (2.2-4.5 mm. forewing length). Head (Figs.
28, 35): frons and vertex smooth-scaled; labial palpus
recurved, apical 2 segments subequal, flattened dorso-
ventrally, basal segment short, segments mostly smooth-
scaled, sometimes roughened; maxillary palpus (Fig. 29) 2-
segmented with minute apical segment and large basal segment; haustellum (Fig. 41) well-developed; pilifer moderate;
ocellus moderate; eye large; antenna (Fig. 47) somewhat
thickened and short. Thorax: smooth-scaled. Forewing (Fig.
57): elongate with falcate apex more or less distinct; pterostigma present; costal margin slightly convex to apex; apex
acutely rounded to falcate indentation of termen; tornus
indistinct, rounded; dorsal margin straight to rounded anal
angle; chorda developed; no vein in cell; Sc to costal margin
before 1/2; R1-R4 to costal margin; R5 to termen; M1-M3
and CuA1 equidistant at end of cell; CuA1 and CuA2 parallel; CuP indistinct; A1+2 with short basal stalk; A3 vestigial.
Hindwing: with sharp convexity at 1/2; apex very acute;
termen very oblique to indistinct tornus; dorsal margin
straight to acute truncate anal angle; cell with vestigial
vein; Sc+R1 to 3/4; Rs to margin before apex, approximate
with M1 at end of cell; M2 near M1 at end of cell; M3
absent; CuA2 divergent from CuA1; A1+2 short; A3 and A4
distinct. Abdomen: often strongly modified around genitalia
as hood-like structure with ventral slit (Figs. 130-131);
coremata usually absent, sometimes present. Male genitalia:
tegumen usually strong, sometimes split by tuba analis;
vinculum strong, fused to tegumen to form continuous ring,
usually with long, thin saccus or saccus reduced or absent,
rarely broad; tuba analis prominent; valva simple or complex,
often split into two separate parts or costal apex and
saccular apex divergent, rarely with accessory appendage
from tegumen base for appearance of second valva; valva

usually setaceous over most of mesal surface; valval coremata absent but coremata sometimes on posterior sternite at
base of ventral split; valval base projected for base of
anellus, often fused with anellus or each fused to produce
entire transtilla; anellus tubular, with aedeagus attached at
tip, often very strongly sclerotized; aedeagus usually very
long and thin, or short, rarely extremely long with ductus
end spiralled; phallobase usually absent; cornutus present as
round tubule or variously modified hooks or teeth, or apparently absent (possibly deciduous).   **Female genitalia:**
ovipositor normal; papilla analis simple, setaceous; apophyses
long and thin, usually subequal; ostium, attached to posterior
edge of sternite 7, usually as protruded tube, sometimes a
cup in a short tube, or with a large ventral sterigma, or a
sclerotized edge; ductus bursa usually membranous, thin and
long, sometimes sclerotized half length from ostium; ductus
seminalis from near mid-point on ductus bursae; bulla seminalis small; corpus bursae elongate-ovate, usually with
spicules on walls; signum absent or present as row of large
teeth or two fused spicule patches.   **Larva** (Figs. 280-288:
D. habecki): head with fronto-clypeus reaching 2/3 to epicranial notch; 2 adfrontal setae; 6 stemmata in semi-circle;
prothorax with L-group bisetose; thoracic legs developed;
mesothorax with single SV seta; abdominal segments with D1
closer together than D2 but segment 6 with D1 slightly more
separated than D2; L1 antero-dorsad of spiracle on abdominal segments; segments 9-10 with sclerotized tergal
plates; tergite 10 with posterior setae very long, stout;
spiracles on produced cylinders, largest on prothorax and
abdominal segment 8; prolegs vestigial; crochets absent.
Pupa (Figs. 294-298: **D. habecki**): elongate with small horn-
like projection on vertex of head, with lateral projecting
spiracles on prothorax; appendages to wing tips; two small
setae on each tergite; no distinct cremaster but ventral and
posterior hook-tipped setae.

      Remarks.- Diploschizia is most closely related to **Gly-
phipterix**, with many species of both genera appearing the
same superficially.   The further reduction in wing venation,
the unusual maxillary palpi, and the additional complexity of
the genitalia in **Diploschizia** indicates an advanced derivation
from **Glyphipterix**.   The maxillary palpi appear to have resulted from secondary fusion of basal segments to produce
the large basal segment found in the genus.   The male genitalia vary considerably among different species of **Diploschizia** but the lack of M3 in the hindwings and the more
coherent variation among the female genitalia of different

species, indicates that the genus is a cohesive unit as herein defined. **Diploschizia impigritella** is the only known species with male abdominal coremata in the genus.

Although the first three species described subsequently have tubular cornuti, it is possible that the last three species described have deciduous cornuti. Females of **Diploschizia kimballi** sometimes have long loose spines in the bursa that appear to be cornuti. Thus far all dissected males of **D. kimballi** have no visible cornuti in the aedeagi but they may all have already mated. Deciduous cornuti are not known for any other glyphipterigids besides the three species possibly having them in this genus.

Biological information is known only for two of the species from America north of Mexico. One (**D. habecki**) feeds inside the seeds of its host in Cyperaceae, moving from seed to seed as each is excavated and until the larva is in final instar. The larva utilizes an empty seed shell of the host as a pupal chamber and forms a simple cocoon therein after chewing a filigreed network on one side for adult eclosion. The pupa is not protruded upon adult emergence. The other known species (**D. impigritella**) has larvae boring in host stems and partially in leaf axils where the pupa is usually later located. In the latter species only a thin layer of stem tissue or leaf axil is left for the adult to break through. The adults congregate around the host plants and are most easily collected from the host or reared, although some have been collected at blacklight.

The genus is confined to the New World with 6 species from North America and 3 described species from the Neotropics, of which the following Neotropical species have been transferred to **Diploschizia** (Heppner, 1981a):

> **Diploschizia glaucophanes** (Meyrick, 1922) (**Glyphip-teryx** [sic])
> **Diploschizia tetratoma** (Meyrick, 1913) (**Glyphip-teryx** [sic])
> **Diploschizia urophora** (Walsingham, 1914) (**Glyphip-teryx** [sic])

In addition to the 4 new species described hereafter, there appear to be a few more described Neotropical species currently in **Glyphipterix** which may be transferred to **Diplo-schizia** after further study and I have seen 4 undescribed species form Mexico that belong in this new genus, bringing the total for the genus to at least 13 species.

Key to Diploschizia Species Based on Maculation

1.   Forewing without dorsal margin crescent (Figs. 122–123) ....**minimella** (p. 140)
Forewing with large dorsal margin crescent..............2

2(1).   Forewing costal margin with 4 white marks..............3
Forewing costal margin with 5 white marks..............4

3(2).   Forewing dorsal margin crescent very broad on margin, with distad very narrow, curved extension (Fig. 124–125) ....**habecki** (p. 142)
Forewing dorsal margin crescent rather uniform tapered from narrow marginal base to distad point (Figs. 120–121) ....**lanista** (p. 138)

4(2).   Forewing with red-brown mid-apical area (Fig. 117)..
....**regia** (p. 145)
Forewing fuscous, without red-brown on wing...........5

5(4).   Apical 1/4 of forewing with small dark fuscous area (Figs. 126–127) (verify determination in genitalia key) ....**impigritella** (p. 146)
Apical 1/4 of forewing with large dark fuscous area (Figs. 118–119) (verify determination in genitalia key) ....**kimballi** (p. 152)

Key to Diploschizia Species Based on Male Genitalia

1.   Saccus reduced, appearing absent, at most very broad and short; aedeagus with distinct cornuti .
....2
Saccus distinct, very narrow, usually long; aedeagus without noticeable cornuti (possibly deciduous) ..
....4

2(1).   Valvae narrowed distally (Fig. 199); aedeagus with 3 long, curved spine-like cornuti (Fig. 200) ....
....**habecki** (p. 142)
Valvae somewhat broad-oblong, not distinctly narrowed distally; aedeagus with single tubule cornutus 3

3(2).   Accessory structures present from tegumen to abdominal sternite, appearing as secondary valvae

(Fig. 197); aedeagus very short; saccus absent ...
....minimella (p. 140)
No accessory structures as described above; aedeagus long; saccus broad and flattened (Fig. 195) ..
.... lanista (p. 138)

4(1).   Valvae with ventral lobe (Fig. 192); valvae fused to strongly sclerotized long anellus (Fig. 193); aedeagus very long, straight (Fig. 194) ....
.... kimballi (p. 152)
Valvae without ventral lobe; valvae attached to very short anellus; aedeagus long but not straight entire length, at least some curvature on anterior end or very curved 5

5(4).   Transtilla with long, setaceous dorsal appendages resembling secondary valvae (Fig. 201) ....
....impigritella (p. 146)
Transtilla without long appendages (Fig. 190)............
....regia (p. 145)

Key to Diploschizia Species Based on Female Genitalia[5]

1.   Ostium bursae a narrow, strongly sclerotized and extended tube from 7th sternite (Fig. 261) 2
Ostium bursae a membranous cup or funnel anterior to small or large 8th sternal modification (Fig. 263) 3

2(1).   Bursa copulatrix with 2 fused spicules in signa (Fig. 262) .... kimballi (p. 152)
Bursa copulatrix without signum (Fig. 268)...............
....impigritella (p. 146)

3(1).   Sternite 8 modified as large sclerotized quadrangular plate with rounded or truncated posterior edge (Fig. 265) ....habecki (p. 142)
Sternite 8 at most with small sclerotized lobes with setae, but no large plate-like structure (Fig. 263)
....minimella (p. 140)

---

[5] The female of Diploschizia regia is unknown; the female of D. lanista is not known sufficiently to be included in the key.

Diploschizia lanista (Meyrick)
(Figs. 16, 120–121, 195–196)

Glyphipteryx [sic] lanista Meyrick, 1918:195; Mc-
    Dunnough, 1939:84; Clarke, 1955:182 [index].
Glyphipteryx [sic] sp., Kimball, 1965:287; Frost,
    1975:40.
Glyphipterix lanista.- Clarke, 1969:68; Heppner,
    1983a:26.
Diploschizia lanista.- Heppner, 1981a:314, 1982a:54.

A small species with a large narrow white crescent on
the forewing dorsal margin and 4 white marks on the apical
costal margin.

Male (Fig. 120).- 2.7–4.0 mm. forewing length. Head:
gray fuscous with small white postero-lateral eye margin;
labial palpus white dorsally, venter with basal segment
white, then 2nd and apical segments each with 2 alternating
bands of black and white; antenna fuscous dorsally. Thorax:
gray fuscous; patagia gray fuscous; venter white with fuscous; legs fuscous with white at joints. Forewing: gray
fuscous over basal 2/3, with apical 1/3 dark fuscous and dark
fuscous borders to all markings, with buff overlaid between
markings near costal margin; dorsal margin with large white
crescent at midwing, point directed toward apex, costal
margin with white oblique bar pointed to tornus, then 3
small white marks near apex, all with mesad silver spot
except 2nd from apex; apex with black spot; silver spot at
falcate indentation and at base of termen; tornus with small
white spot with mesad silver bar; fringe fuscous, white
distally except all white at falcate indentation; venter
fuscous with dorsal white marks faintly repeated except
more distinct apical marks. Hindwing: fuscous; fringe
fuscous; venter whitish fuscous. Abdomen: fuscous with
silvery scales on posterior of each segment; venter same but
mostly white on posterior of each segment; 8th abdominal
segment modified with acute sclerotized point on each side
of ventral split; coremata absent. Genitalia (Fig. 195): tuba
analis moderate, broad; tegumen an elongate inverted U-
shape, stout; vinculum small, merging to broad spatulate

saccus; valva elongate, wider near apex, termen rounded, setaceous; short tubular anellus somewhat membranous; transtilla arms free; aedeagus (Fig. 196) short (3/4 valval length), with apical end narrower than anterior end; cornutus a long tubule; vesica with spicules; ductus ejaculatorius with small hood somewhat distant from aedeagus (8 preparations examined).

Female (Fig. 121).- 3.0 mm. forewing length. Similar to male. Genitalia: ovipositor short; papilla analis small, setaceous; apophyses very thin, long, anterior and posterior pair subequal (1 preparation examined. The only available female specimen was received with the abdomen partially damaged by dermistids and only those parts of the genitalia noted above remained for study).

Type.- Holotype male: Southern Pines, [Moore Co.], North Carolina, May 1916, Parish (BMNH).

Additional specimens (50 males, 1 female).- Florida.- Alachua Co.: Gainesville, 8 Jul 1966 (1 male), L. O'Berry (CPK). Dade Co.: Royal Palm St. Park [=Royal Palm Hammock, Everglades Natl. Park], Jan 1930 (1 male), F. M. Jones (USNM). Escambia Co.: Pensacola, 13 Sep 1961 (1 male, CPK), 30 Sep 1961 (1 female, FSCA), S. O. Hills. Highlands Co.: Archbold Biol. Sta., [10 mi. (=16 km.) S. Lake Placid], 1 Jan 1965 (1 male), 10 Jan 1969 (1 male), 29 Dec 1968 (1 male), S. W. Frost (FEM); 1-7 May 1964 (1 male), 16-22 May 1964 (1 male), R. W. Hodges (USNM); 4 May 1975 (1 male), at blacklight, J. B. Heppner (JBH). Okaloosa Co.: Shalimar, 7 Nov 1966 (2 males), H. O. Hilton (CPK). Orange Co.: Winter Park, May 1946 (1 male), A. B. Klots (AMNH). Louisiana.- Natchitoches Co.: 4 mi. NW. Gorum, 4 Apr 1970 (1 male), G. Strickland (USNM). North Carolina.- Moore Co.: Southern Pines, 1918 (2 males, BMNH; 1 male, USNM), Parish; [Southern Pines?], [no date] (1 male, ANSP; 1 male, LACM; 8 males, USNM); May (3 males), 1-7 Aug (1 male), 8-15 Aug (2 males), 16-23 Aug (4 males), 24-31 Aug (2 males), 8-15 Sep (1 male), 16-23 Sep (2 males), (USNM).

Distribution (Fig. 16).- Collection records are from North Carolina, Florida, and Louisiana.

Flight period.- April to May, July to September (North Carolina); January, May, July, September, November, and December (Florida); April (Louisiana).

Hosts.- Unknown.

Biology.- Unknown.

Remarks.- Specimens vary in the extent of forewing buff scaling and the length of the 4th white mark from the apex along the costal margin, which is sometimes reduced but then has more silver scaling.

Diploschizia lanista is widespread in the Southeast. Virtually all available specimens are males and the single known female was damaged, as noted above, and did not provide many genital characters for comparison with females of other species. The species appears to be the least apomorphic of the genus, as indicated by the simplified male genitalia but which show some similarity to the next two species. It appears to be at the beginning of the progression to the more complex genitalia of the other species in the genus.

Diploschizia minimella Heppner, 1981
(Figs. 17, 122-123, 197-198, 263-264)

Diploschizia minimella Heppner, 1981a:315, 1982a:54.

This is one of the smallest species of glyphipterigid and has no midwing crescent on the dorsal margin of the forewing as is common in the genus.

Male (Fig. 122).- 2.2-2.9 mm. forewing length. **Head:** fuscous with bronze shine, without lateral white eye margin; labial palpus dorsally white with some fuscous on apical segment, venter same but more fuscous on 2nd and apical segments, somewhat rough-scaled; antenna dorsally dark fuscous. **Thorax:** fuscous; patagia fuscous with bronze shine; venter silvery white with some fuscous; legs fuscous with white at joints. **Forewing:** pale fuscous ground color with bronze shine, overlaid on apical 1/3 with buff except dark fuscous borders of markings; costal margin with 4 white marks from beyond midwing to apex with basal mark longest and oblique, pointed to tornus, each with more or less distinct silver spot mesad; dorsal margin with white bar basad of tornus with silver spot mesad; black spot at apex; silver spot near termen and at falcate indentation; fringe fuscous; white distally with all white at falcate indentation; venter fuscous with costal marks repeated and with apical black spot. **Hindwing:** fuscous; fringe and venter fuscous. **Abdomen:** fuscous with silvery scales on posterior of each segment; venter mostly white; 8th abdominal segment modified as hood for male genitalia with ventral split and setaceous ventral ends. **Genitalia (Fig. 197):** tuba analis broad; tegumen constricted between dorsum and valval bases with fused sclerotization of intersegmental membrane convergent to transtilla from valval base; vinculum subquadrate with concave saccus edge; saccus absent; elongate setaceous process divergent from tegumen base with web-like membrane attached to tegumen and 8th abdominal pleurite and vinculum; valva elongate, setaceous, with acute termen and convex dorsal margin near base, then angled straight to apex, with saccular margin similar but concave; transtilla fused as border to short anellus; aedeagus (Fig. 198) short (2/3 valval length), narrow; phallobase absent; cornutus a large constricted tube; vesica without spicules; ductus ejaculatorius with moderate hood area near aedeagus (2 preparations examined).

Female (Fig. 123).- 2.3-2.7 mm. forewing length. Similar to male. Genitalia (Fig. 264): ovipositor stout; papilla analis small, setaceous; apophyses long, thin, subequal; 8th sternite with invaginated sclerotized edge with several setae (sometimes reduced to 2 setae); ostium bursae (Fig. 263) a membranous circle and short funnel with sclerotized ring; ductus bursae short, membranous, merging to very elongate, moderate bursa copulatrix; ductus seminalis from ductus bursae; corpus bursae without distinct signum but with several dozen strong short spines (2 preparations examined).

Types.- Holotype male: Archbold Biol. Sta., [10 mi. (=16 km.) S.] Lake Placid, [Highlands Co.], Florida, 16-22 May 1964, R. W. Hodges (USNM). Paratypes (2 males, 5 females): Florida.- Highlands Co.: same locality as holotype, 29 Mar 1959 (1 female), R. W. Hodges (USNM); 15-31 Jul 1948 (1 female), A. B. Klots (AMNH); Lake Placid, 30 Apr 1964 (1 female), R. W. Hodges (USNM); Sebring, 13 Aug 1942 (1 female), 25 Aug 1942 (1 male), C. T. Parsons (MCZ). Orange Co.: Orlando, 18 Feb (1 male), G. G. Ainslie (USNM). Santa Rosa Co.: Munson Cpgd., 2 mi. [=3.2 km.] E. Munson, 8 Jun 1975 (1 female), K. W. Knopf (FSCA).

Distribution (Fig. 17).- Known only from Florida.

Flight period,- February to May; June to August.

Hosts.- Unknown.

Biology.- Unknown.

Remarks.- This species is similar to **Diploschizia lanista** but typically is smaller and has the forewing dorsal margin crescent absent. The male genitalia have a unique development of the last abdominal segment reaching from the tegumen to the end of the abdomen, forming what appear to be secondary valvae.

Diploschizia habecki Heppner, 1981
(Figs. 17, 124–125, 199–200, 265–266)

Diploschizia habecki Heppner, 1981a:317, 1982a:54.

The very broad base of the forewing dorsal margin crescent, with the slender distal point, is characteristic of the
species.

Male (Fig. 124).- 2.6–3.4 mm. forewing length. **Head:**
fuscous with reduced white lateral eye margin; labial palpus
dorsally white with some fuscous; venter with basal segment
white, 2nd segment with 2 alternating bands of black and
buff-white with apex black, with buff-white laterally; antenna dark fuscous dorsally. **Thorax:** fuscous; patagia
fuscous; venter white; legs fuscous with white on femur and
joints. **Forewing:** ground color dark fuscous with bronze
shine and brown-buff overlaid scaling on anal 1/4 and apical
half except for dark fuscous borders of markings; dorsal
margin with white crescent at midwing with broad semi-
circular base nearer 1/3 from base and distal very slender
extension curved to midwing, with some buff on mark along
margin; costal margin with 4 white bars from near 2/3 to
apex, with basal mark oblique and longest, all with silver
spot or line mesad; small dark fuscous spot at end of cell
and larger spot mid-apically; apex with black spot; silver
spot at falcate indentation; 2 silver spots along termen;
tornus with small white spot with silver bar mesad followed
further disto-mesad by another silver spot; fringe fuscous,
white distally with all white at falcate indentation; venter
fuscous with dorsal white marks faintly repeated except
distinct apical marks. **Hindwing:** fuscous; fringe and venter
fuscous. **Abdomen:** fuscous with silvery scales on posterior
of each segment; venter mostly white; 8th segment modified
as covering for male genitalia with ventral split and acute,
setaceous ventral ends; coremata absent. **Genitalia** (Fig.
199): tuba analis broad, short; tegumen narrow, merging into
lateral edges of subquadrate vinculum to form ovate fringe-
like structure; saccus very reduced; valva very elongate,
with broad base tapering to truncate narrow distal end with
setae and several spines; valval base extended dorsally as
transtilla but ends free; anellus short, little sclerotized;
aedeagus (Fig. 200) short (half valval length), narrow; cornutus as 3 very large curved, divergent spines, with 2-3
barbs apically on two and one spine smooth; vesica without

spicules; ductus ejaculatorius with small hood near aedeagus; phallobase reduced (6 preparations examined).

Female (Fig. 125).- 2.4-3.3 mm. forewing length. Similar to male. Genitalia (Fig. 266): ovipositor short; papilla analis small, setaceous; apophyses long, thin, posterior pair 1/3 longer than anterior pair; 8th sternite modified as sclerotized rectangular shield with rounded or truncate, sharp posterior edge; ostium bursae (Fig. 265) anterior to and invaginated from shield of 8th sternite, a membranous funnel; ductus bursae a short membranous tube, thin; ductus seminalis emergent from ductus bursae near bursa; corpus bursae moderate; elongate-ovate with numerous spicules but no distinct signum (4 preparations examined).

Larva (Figs. 280-288).- Integument rugose, unicolorous white with brown tergal plates and pinacula; head amber with A2 close to A3, distant to A1; prothorax with L1 approximate to L2; SV1 approximate to SV2; abdominal segment 1 with D1 closer together than D2, segment 6 with D2 slightly closer together than D1; segments 9-10 with sclerotized tergites, with 10th tergite large and with 5 very large posterior setae; spiracle on produced cylindrical structures, longest on prothorax and abdominal segment 8, the latter with produced bulbous base.

Pupa (Figs. 294-298).- Amber colored; large projecting spiracle from prothorax dorso-laterally; small pointed projection on vertex of head; abdomen slender, elongate, with pair of setae on each of the tergites; cremaster absent except as several hook-tipped setae.

Types.- Holotype male: Bivens Arm lake, 3 mi [=4.8 km.] SW. Gainesville, Alachua Co., Florida, 27 Oct 1974, reared ex **Rhynchospora corniculata**, emerged 5 Nov 1974, J. B. Heppner (FSCA). Paratypes (100 males, 91 females): **Florida.**- Alachua Co.: same locality as holotype (all reared from **Rhynchospora corniculata**), 1 Jul 1974 (2 males, emerged 10 Jul; 1 female, 15 Jul; 9 males, 10 females, 17 Jul), D. H. Habeck (FSCA); 1 Jul 1976 (2 males, 1 female, emerged 10 Jul), P. A. Travis (FSCA); 3 Jul 1973 (1 male, emerged 16 Jul), D. H. Habeck (FSCA); 8 Aug 1972 (1 male, 3 females, emerged 15 Aug; 1 male, 1 female, 23 Aug; 1 female, 24 Aug; 1 male, 27 Aug; 2 males, 28 Aug; 1 male, 6 females, 30 Aug; 9 males, 4 females [no dated emergence]), D. H. Habeck (FSCA); 16 Aug 1972 (1 male, 1 female, emerged 30 Aug; 2 males, 1 female, 1 Sep), D. H. Habeck (FSCA); 16 Aug 1972 (4 males, 3 females, emerged 29 Aug; 2 males, 3 females, 1 Sep), J. B. Heppner (FSCA); 4 Sep 1972 (3 males, emerged, no dated emergence), D. H. Habeck

(FSCA); 12 Sep 1974 (3 males, emerged 18 Sep; 2 males, 4 females, 24 Sep; 5 males, Nov), J. B. Heppner (JBH); 2 Oct 1975 (14 males, 7 females, emerged 18-20 Oct), J. B. Heppner (JBH); 9 Oct 1973 (1 female, emerged 16 Oct; 1 male, 2 females, 23 Oct), D. H. Habeck (FSCA); 27 Oct 1974 (4 males, 1 female, emerged 5 Nov), J. B. Heppner (JBH); Archer Road Lab, 3 mi. [=4.8 km.] SW. Gainesville, 30 Aug 1975 (1 female), J. B. Heppner (JBH); Gainesville, 24-28 Dec 1975 (1 female), W. H. Pierce (FSCA).   Highlands Co.: Highlands Hammock St. Park, 4 May 1974 (1 male, emerged 11 May; 1 male, 12 May; 1 male, 15 May; 1 male, 16 May; 2 females, 19 May; 3 females, 20 May; 6 males, 6 females, 21 May; 3 females, 22 May; 4 males, 4 females, 23 May; 9 males, 6 females, 24 May; 5 males, 5 females, 25 May; 3 males, 6 females, 26 May; 1 male, 29 May), reared ex **Rhynchospora corniculata**, J. B. Heppner (JBH).   Lake Co.: Alexander Springs Cpgd, 6 mi. [=9.6 km.] S. Astor Park, 21 Apr 1975 (1 female), J. B. Heppner (JBH). **Georgia.**- Thomas Co.: Ochlockonee R., 6 mi. [=9.6 km.] W. Thomasville, 20 Oct 1975 (1 male, emerged 29 Oct; 1 female, 4 Nov; 1 female, 7 Nov), reared ex **Rhynchospora corniculata**, J. B. Heppner (JBH). (Paratypes to BMNH, CNC, UCB, USNM).

Distribution (Fig. 17),- Southern Georgia to central Florida.

Flight period.- April to May; July to December (probably also June).

Hosts.- **Rhynchospora corniculata** (Lamarck) Gray (Cyperaceae).

Biology.- The larvae are seed borers of the single known host plant. Larval activity can be seen by the frass deposited on the exterior of seeds with occupying larvae (one larva per seed). After several seeds are consumed, the last seed to be occupied is used as the pupal chamber by the construction of a filigreed network on one side of the seed; the larva chews the seed wall to form the filigreed network. At ecdysis the pupa is not protruded but the adult emerges through this filigreed network. The average development period from young larva to adult is about 10-14 days. Typical habitats are mainly marshy areas near creeks or ponds where the host plant is often common.

Remarks.- Variation among the available specimens mainly involves slight alterations in the size of forewing markings and in the genitalia with variations in the shape of the 8th sternal plate, either rounded, as in the female illustrated (Fig. 265), or very truncated with a straight edge. it is possible that the females use this sternal plate as a piercing organ for egg deposition, inasmuch as the ovipositor is relatively unsclerotized otherwise.

The genitalia show some relationships to both **Diploschizia lanista** and **D. minimella** but are otherwise distinct in the genus in many features, especially the very large and multiple cornuti. There are two species, **Diploschizia urophora** and **D. tetratoma**, both occurring in Mexico (the latter species also in Brazil), which have valvae very much like **D. habecki**. These species do not have such unusual cornuti but are otherwise similar in the type of male genitalia. Both have narrow forewing crescent marks.

Diploschizia regia Heppner, 1981
(Figs. 17, 117, 190-191)

Diploschizia regia Heppner, 1981a:320, 1982a:54.

A small light-colored species with 5 costal margin white marks, a red-brown mid-apical area, and distinctive genitalia.

Male (Fig. 117),- 2.8 mm. forewing length. **Head:** gray-buff; labial palpus dorsally white, venter white with buff and fuscous apically; antenna dorsally fuscous. **Thorax:** silvery gray; patagia silvery gray; venter silvery white; legs mostly white with some fuscous between joints. **Forewing:** silvery gray ground color with buff on apical 1/3 becoming red-brown on apical center except fuscous borders to all markings; dorsal margin with large white crescent at midwing; costal margin with 5 white bars from near midwing to apex, with apical 3 small and basal 2 larger, with 4th from apex extended to tornus as curved silver fascia; apex with black spot; silver spot at tornus and at falcate indentation; fringe fuscous, distally white except all white at falcate indentation; venter fuscous with apical marks distinct and similar to dorsal marks. **Hindwing:** shining fuscous; fringe and venter fuscous. **Abdomen:** fuscous with silvery scales on posterior of each segment; venter mostly white; 8th segment modified in male with ventral split and setaceous ends; coremata absent. **Genitalia** (Fig. 190): tuba analis as broad as tegumen space; tegumen split dorsally and separated, extended from triangular vinculum; saccus long, narrow; vinculum with small stub near valval base each side; valva elongate-oblong, setaceous, with somewhat acute point, with dorsal base extended as long transtilla with ends fused; long tubular anellus with distal setaceous end, fused to valva

ventral base and having ventral stub-like projection; aedeagus
(Fig. 191) very long (somewhat longer than distance from
tuba analis to end of saccus), thin, relatively straight;
cornutus absent (possibly deciduous); vesica without spicules;
ductus ejaculatorius with hood distant from aedeagus (1
preparation examined).

Female.- Unknown.

Type.- Holotype male: Royal Palm St. Park [= Royal
Palm Hammock, Everglades Natl. Park, Dade Co.], Florida,
Jan 1930, F. M. Jones (USNM).

Distribution (Fig. 17).- Known only from southern Florida.

Flight period.- January

Hosts.- Unknown.

Biology.- Unknown.

Remarks.- This species is easily distinguished from other
Diploschizia by its lighter coloration, the 5 costal marks, and
the distinctive genitalia. This species and the following two,
and similar Neotropical species, all have very long aedeagi
and lack cornuti (but possibly have deciduous cornuti). The
corresponding females all have a projected ostium arrangement with a sclerotized portion on the ductus bursae, thus,
the female of D. **regia** probably has genitalia similar to
these other females.

Diploschizia impigritella (Clemens)
(Figs. 18, 126-127, 130, 201-202, 267-268)

Glyphipteryx [sic] **impigritella** Clemens, 1863:9;
Stainton, 1870:232; 1872:xi; Zeller, 1877:403;
Chambers, 1878b:148; Riley, 1891:104; Dyar,
1900:84; Smith, [1900]:479; Busck, 1903:211;
Dyar, [1903]:493; Kearfott, 1903:108; Anderson,
1904:53; Busck, 1904:750; Engel, 1908:121; Morse,
1910:553; Meyrick, 1913b:42; 1914c:30; Walsingham, 1914:300; Barnes & McDunnough, 1917:
182; Forbes, 1923:356; Leonard, 1928:554;
Brimley, 1938:310; Procter, 1938:266; McDunnough, 1939:84; Procter, 1946:318; Glick, 1965:
134; Kimball, 1965:287.

Glyphipteryx [sic] **exoptatella** Chambers, 1875a:234;
1875b:293; 1878b:148; Riley, 1891:104; Busck,
1903:211; Kearfott, 1903:108 [in syn.].

Glyphipteryx [sic] sp.- Frost, 1964:153.

Diploschizia impigritella.- Heppner, 1981a:321, 1982a:
54, 1982b:249; Covell, 1984:432.
Glyphipterix impigritella.- Heppner, 1983a:26.

A very widespread species distinguished by the genitalia
from superficially very similar species both of the Nearctic
and Neotropical regions.

Male (Fig. 126).- 2.7–4.3 mm. forewing length. **Head:**
fuscous with reduced white line near antennal base; labial
palpus white dorsally, venter white except for medial fuscous
line from mid-2nd segment to apex of apical segment; antenna dorsally fuscous. **Thorax:** fuscous; patagia fuscous;
venter white and fuscous; legs fuscous with white at joints.
Forewing: fuscous ground color overlaid with yellow-buff on
apical half except for fuscous borders of markings; dorsal
margin with large white crescent from basal 1/3 to midwing,
pointed toward apex; costal margin with 5 white bars from
half to apex with basal mark longest white and oblique to
center of wing, other marks with silver spot each mesad or
reduced silver; apex with black spot; large fuscous area mid-
apically; silver spot at falcate indentation; silver spot near
tornus along termen and a small white spot with silver mesad at tornus; fringe fuscous with white distally except all
white at falcate indentation; venter fuscous with dorsal
marks faintly repeated except distinct apical marks. **Hindwing:** fuscous; fringe fuscous; venter silvery fuscous.
Abdomen: fuscous with silvery scale row on posterior of
each segment; venter mostly white; coremata present; 8th
segment (Fig. 130) of male abdomen modified as hood for
genitalia with venter split and lateral posterior ends pointed
somewhat and setaceous on internal edge. **Genitalia** (Fig.
201): tuba analis short, narrow; tegumen elongate, stout
with wider dorsum, merging to narrow truncated vinculum;
saccus very long (longer than distance from tuba analis to
end of vinculum), narrow; valva short, setaceous, with broad
base tapering abruptly after saccular convexity to narrower
distal end with termen having several stout, short spines;
dorsal base of valva very stout, elongated as stub centrally
fused as transtilla, with very long setaceous and sclerotized
process divergent from transtilla with apical spines and resembling a second pair of valvae; anellus a short, strongly
sclerotized tube fused to transtilla; aedeagus (Fig. 202) very
long (1.25 times length of saccus), narrow, curved; phallobase
absent; cornutus absent (possibly deciduous); vesica without

spicules; ductus ejaculatorius with hood relatively close to
aedeagus (16 preparations examined).

Female (Fig. 127).- 4.0-4.5 mm. forewing length. Similar
to male. Genitalia (Fig.268): ovipositor short; papilla analis
small, setaceous; apophyses thin with posterior pair long,
twice length of short anterior pair; 8th sternite unmodified;
ostium bursae (Fig. 267) a small opening on a strongly
sclerotized projected tube encompassing part of the
sclerotized anterior of the ductus bursae, with ostium
opening ventrad with two apical ventro-lateral ridges; ductus
bursae thin, with anterior half membranous; ductus seminalis
emergent from bursa immediately proximal to ductus bursae
entrance; corpus bursae moderate, elongate-ovate; signum
absent but with numerous spicules on posterior end (19
preparations examined).

Types.- Holotype female (impigritella): "178" [Easton,
Northhampton Co., Pennsylvania, no date, B. Clemens?]
(ANSP, type 7325). Holotype (no abdomen) (exoptatella):
[Covington?, Kenton Co.], "Kentucky Chambers" (MCZ, type
1564).

Additional specimens (149 males, 80 females).-
Arkansas.- Montgomery Co.: Fiddlers Cr., 11 Jun 1975 (1
female), H. N. Greenbaum (JBH). Washington Co.: 4 Jul 1966
(1 male), R. L. Brown (USNM); Devil's Den St. Park, 22 May
1966 (1 male, 1 female), 24 May 1966 (1 male), 28 May 1966
(1 female), 11 June 1966 (1 male), 24 Jun 1966 (1 male), 25
Jun 1966 (5 males, 1 female), 27 Jun 1966 (1 male), 4 Jul
1966 (1 male), 9 Jul 1966 (1 male), 11 Jul 1966 (1 male), R.
W. Hodges (USNM). California.- Siskiyou Co.: Brown's Lake,
SW. Mt. Shasta City, 13-14 Jun 1974 (1 female), J. A. Powell
(UCB); Mt. Shasta City, 18 Jul 1958 (2 males), J. A. Powell
(UCB). Trinity Co.: Buttercreek Meadows, 8 mi. [=12.8
km.] W. Hayfork, 20 May 1973 (1 female), 3750' [=1140 m.],
R. Dietz (UCB). Delaware.- Newcastle Co.: Newark, 24 Jul
1970 (1 male), 15 Jul 1968 (1 male), 16 Jul 1968 (1 female),
2 Aug 1969 (1 male), 5 Aug 1968 (1 female), 14 Aug 1968 (1
male); 17 Aug 1968 (1 male), 21 Aug 1968 (1 male), D. F.
Bray (DFB). District of Columbia.- Washington, 26 May
1908 (1 female), C. R. Ely (LACM); Jun 1902 (1 male,
LACM; 1 male, USNM), 2 Jun 1902 (1 female, BMNH), A.
Busck; 26 Jun 1963 (1 male), D. C. & K. A. Rentz (CAS).
Florida.- Alachua Co.: Archer Road Lab, 3 mi [=4.8 km.]
SW. Gainesville, 3 Apr 1976 (1 male), emerged ex Juncus?,
12 Apr, J. B. Heppner (JBH); 4 Apr 1976 (1 female), 2 May
1976 (1 female), 3 May 1976 (1 female), 3 Aug 1975 (1
male), at blacklight, J. B. Heppner (JBH); Austin Cary

Forest, 6 mi [=9.6 km.] NE. Gainesville, 14-15 Apr 1975 (1 male, 1 female), ex Malaise trap, G. B. Fairchild (FSCA); Gainesville, 8-12 Mar 1976 (1 female), 20 Mar 1976 (1 male), 14 Apr 1976 (1 male), ex Malaise trap, W. H. Pierce (FSCA); 27 Mar 1975 (1 male), ex Malaise trap, H. N. Greenbaum (JBH); Univ. Fla. Hort(iculture) Unit, 9 mi. [=14.4 km.]) NW. Gainesville, 26-27 Mar 1975 (1 female), ex Malaise trap, G. B. Fairchild (FSCA). Baker Co.: 4 mi. [=6.4 km.] SW. Macclenny, 16 May 1975 (2 females), on flowers Pyracantha, J. B. Heppner (FSCA). Glades Co.: Fisheating Cr., Palmdale, 7-10 May 1964 (6 males, 2 females), R. W. Hodges (USNM). Highlands Co.: Archbold Biol. Sta., 10 mi [=16 km.] S. Lake Placid, 10 Mar 1963 (1 male), 11 Mar 1970 (1 male), 30 Mar 1970 (1 male), 22 Apr 1969 (1 female), 24 Apr 1968 (1 female), 24 Apr 1969 (1 male), S. W. Frost (FEM); 3 Apr 1959 (1 male), 1-7 May 1964 (1 male, 1 female), R. W. Hodges (USNM); Lake Placid, 30 Apr 1964 (2 females), R. W. Hodges (USNM). Manatee Co.: Gulf Coast Exp. Sta., Bradenton, 19 Sep 1955 (1 male), E. G. Kelsheimer (CPK). Polk Co.: Peace R., 5 mi [=8 km.] S. Ft. Meade, 22 Apr 1975 (1 female), at blacklight, J. B. Heppner (JBH). Georgia.- Clarke Co.: May 1929 (1 male), Richards (AMNH). Illinois.- Putnam Co.: 15 Sep 1936 (1 male), M. O. Glenn (USNM). Iowa.- Story Co.: Ames, 28 Jun 1961 (1 male), W. S. Craig (ISU). Kansas.- Pottawotomie Co.: Onaga, [no date] (1 male), (MCZ). Louisiana.- East Baton Rouge Co.: Baton Rouge, 28 Mar 1971 (1 male), 18 Apr 1971 (1 male), 19 Apr 1971 (1 male), 21 Apr 1971 (1 male), 22 Apr 1970 (1 male), 23 Apr 1971 (2 males), 24 Apr 1971 (2 males), 25 Apr 1971 (3 males, 3 females), 27 Apr 1971 (3 males), 30 Aug 1969 (1 female), 23 Oct 1971 (1 female), G. Strickland (USNM). Madison Co.: Tallulah, [no date] (1 male), (USNM). Orleans Co.: New Orleans, 20 Aug 1974 (1 female), 16 Sep 1974 (1 female), V. A. Brou (VAB). Maryland.- Prince Georges Co.: Oxon Hill, 29 Jun 1972 (2 males), 2 Jul 1972 (3 males), G. F. Hevel (USNM). Michigan.- Midland Co.: 5 Sep 1959 (1 male), R. R. Driesbach (CNC). Mississippi.- Franklin Co.: Clear Springs Cpgd., 10 mi [=16 km.] SW. Meadville, 20 Apr 1976 (1 female), at blacklight, J. B. Heppner (JBH). Hinds Co.: Clinton, 18 May 1971 (1 male, 1 female), B. Mather (BM); Jackson, 20 Jul 1974 (1 female), B. Mather (BM). Washington Co.: Stoneville, 29 Apr 1975 (1 male, 1 female), 30 Apr 1975 (1 male), 4 Jun 1975 (1 male, emerged 10 Jun), 6 Jun 1975 (1 male emerged 18 Jun), 27 Sep 1973 (1 male, emerged 13 Oct; 1 male, 16 oct; 1 male, 18 Oct; 5 males, 1 female, 19 Oct), reared ex Cyperus rotundus, K. E. Frick

(MSU). **Missouri.**- St. Louis Co.: St. Louis, 10 Jun 1905 (1 female), McElhose (FMNH); Webster Groves, 7 Jun 1919 (1 male), Satterthwalt (USNM). **Nevada.**- Nye Co.: Currant Cr. Cpgd., 20 Jul 1968 (1 male), Opler, Powell, & Scott (UCB). **New Hampshire.**- Rockingham Co.: Hampton, 10 Jun 1905 (1 male), S. A. Shaw (USNM). **New Jersey.**- Burlington Co.: New Lisbon, 1 Jul 1936 (4 males), E. P. Darlington (ANSP). Cape May Co.: 5M Beach, 3 Jul (1 female), F. Haimbach (ANSP). Middlesex Co.: New Brunswick, 3 Jun 1929 (1 female), (AMNH). Ocean Co.: Lakehurst, 2 Jun 1962 (1 male), R. W. Hodges (USNM). **New York.**- "N. Y." (1 male), Beutenmüller (USNM). Monroe Co.: 29 Jul 1949 (1 female), 30 Jul 1949 (1 male), 1 Aug 1949 (1 male), 2 Aug 1948 (2 males), 22 Aug 1949 (1 male), 28 Aug 1949 (2 males), C. P. Kimball (CPK). Tompkins Co.: Ithaca, 12 Sep 1957 (1 female), D. R. Davis (USNM). **North Carolina.**- Macon Co.: Highlands, Apr-May 1938 (1 female), 3-5000' [=915-1520 m.], R. C. Shannon (USNM); 26 Aug 1958 (1 male), 3865' [=1180 m.], R. W. Hodges (USNM). Yancey Co.: Black Mts., 25 May (1 male, 1 female, USNM), 8 Jun (1 female, AMNH), 12 Jun (1 male, 1 female, AMNH), 21 Jun (1 female, AMNH). **Ohio.**- Hamilton Co.: Cincinnati, 7 Jun 1914 (1 female), 9 Jun 1907 (1 female), 28 Jun 1919 (1 male), 14 Jul 1909 (1 female), 11 Sep 1907 (1 female), A. F. Braun (USNM); 17 Jul 1907 (1 female), 30 Jul 1907 (2 males), 14 Aug 1907 (1 female), 30 Aug 1903 (1 male), 31 Aug 1907 (1 female), A. F.Braun (ANSP). **Oklahoma.**- Oklahoma City, 6 Aug 1955 (1 male), 30 Aug 1955 (1 female, D. R. Davis (USNM). **Oregon.**- Douglas Co.: 15 mi. [=24 km.] SW. Diamond Lake, 23 Jul 1966 (1 female), P. Rude (UCB). **Pennsylvania.**- Beaver Co.: New Brighton, 14 Jun 1907 (1 male), (USNM). Luzerne Co.: Hazelton, 18 Jul 1895 (1 female), Dietz (BMNH); 6 May 1903 (1 male), 3 Jun 1904 (1 male), 5 Jun 1897 (1 male), 19 Jun 1897 (1 male), 29 Jul 1897 (1 male), 2 Aug 1905 (2 males), 27 Aug 1919 (1 male), Dietz (MCZ); 2 Jun 1897 (1 male), 2 Aug 1905 (1 male), Dietz (USNM); 2 Jun 1898 (1 male), 22 Jul 1897 (1 male), Dietz (LACM). Northampton Co.: [Easton?], 1872 (1 male), B. Clemens (BMNH). **Texas.**- [Dallas], 9 Mar (1 male), Boll (ex C. V. Riley Coll.) (USNM). Bell Co.: Belton Resv., 6 May 1970 (1 male), A. & M. E. Blanchard (AB). Howard Co.: 15mi. [=24 km] NW. Big Spring, 13 Jun 1963 (1 female), D. C. & K. A. Rentz (CAS). **Virginia.**- Arlington Co.: Arlington, 13 Jul 1906 (1 male, 1 female), emerged ex "chufa" (USNM). Fairfax Co.: Alexandria (Rose Hill), 21 Apr 1976 (1 male), 17 Jun 1976 (1 male), 29 Jun 1976 (2 females), 30 Jul 1976 (2 males), at blacklight, P. A. Opler (USNM); Falls Church, 6 Jun 1962 (1 male), O. S. Flint (USNM). **West Virginia**.- Preston Co.: Aurora, 2 Sep 1904 (1 female), O. Heidemann (LACM). **Canada**.- **British Columbia**.- Mission City, 9 Jun 1953 (1 female), 14 Jun 1953 (1 male), 18 Jun 1953 (1 male), 26 Jun 1953 (1 female), W. R. M. Mason (CNC). **Manitoba**.- Farnworth Lake, Churchill, 14 Jul 1952 (1 female), J. G. Chillcott (CNC). **New Brunswick**.- Northumberland Co.: Tabusintac, 29 Jul 1939 (1 female), J. H. McDunnough (CNC). Restigouche Co.: Jacquet River, 24 Jun 1941 (1 male), J. H. McDunnough (CNC). **Newfoundland**.- Bonavista Bay, NW arm, 3 Jul 1961 (1 male); Terra Nova Natl. Park, 6 Jul 1961 (1 male), 7 Jul 1961 (2 males, 1 female), C. P. Alexander (USNM). **Nova Scotia**.- Colchester Co.: Economy Point, 26 Jun 1957 (3 males), D. C. Ferguson (NSM). Shelburne Co.: Sable Is. (west end), 5 Jul 1967 (1 male), 6 Jul 1967 (3 males, 1 female), 8 Jul 1967 (2 males, 1 female), 11 Jul 1967 (2 females), 13 Jul 1967 (4 males, 2 females), D. M. Wood (CNC). **Ontario**.- Carleton Co.: Ottawa, 27 Jul 1934 (1 male), C. H. Young (CNC). Thunder Bay Co.: Thunder Bay, Jul 1945 (1 female), H. S. Parish (USNM). **Quebec**.- Montcalm Co.: Escalier Lk., Mt. Tremblant Prov. Park, 16 Aug 1973 (2 females), J. B. Heppner (JBH).

Distribution (Fig. 18).- Widely distributed in eastern North America, from Newfoundland to Florida, west to the Great Plains from Manitoba to central Texas; also from British Columbia to northern California; Nevada.

Flight period.- June to August (Canada and northern United States); May to September (middle latitudes of United States, including Pacific Coast); March to September (Southeast).

Hosts.- **Cyperus esculentus** Linnaeus and **Cyperus rotundus** Linnaeus (Cyperaceae). The **Juncus** record from Florida may be erroneous.

Biology.- Larvae bore in the stems of the hosts and pupate in leaf axils in most cases (K. E. Frick, pers. comm.). Field collected late instar larvae develop to adults in 16-22 days in Mississippi (label data). The two known host plants in the genus **Cyperus** appear to be closely related since they are placed next to each other in floras in which both species are treated, among others. The only flower visitation record is for ornamental firethorn, **Pyracantha** sp. (Rosaceae), in Florida.

Remarks.- This is the most widespread and most commonly encountered glyphipterigid in North America. It is relatively uniform throughout its range and varies only in minor shape or size variations of forewing spots. The male genitalia are among the most complex in the genus and in the family but some development to this complexity is evident in two related species from Mexico, both undescribed.

Diploschizia impigritella has been reported from the Neotropics (Walsingham, 1914) but these reports refer to undescribed species superficially almost identical but with very distinct genitalia. Our species is most closely related to the undescribed species from Mexico and beyond these is further related to Diploschizia kimballi Heppner. Relationships are also evident, especially through the female genitalia, to the Neotropical D. tetratoma and, thus, to its related species, including D. habecki and D. kimballi.

Florida specimens are very difficult to distinguish from D. kimballi and should be dissected to verify their identity. Generally D. impigritella is lighter on the apical part of the forewing than D. kimballi.

Diploschizia kimballi Heppner, 1981
(Figs. 19, 118–119, 131, 192–194, 260–262)

Diploschizia kimballi Heppner, 1981a:325, 1982a:54.

This Floridian species is virtually identical superficially to Diploschizia impigritella, being on average only somewhat darker on the apical quarter of the forewing and having the midwing crescent somewhat more slender. The genitalia are very distinct.

Male (Fig. 118).- 3.0–4.0 mm. forewing length. Head: fuscous, with small white line by antennal base; labial palpus white dorsally, venter same but with fuscous central line from middle of 2nd segment to apex of apical segment; antenna dorsally fuscous. Thorax: fuscous; patagia fuscous; venter white with some fuscous; legs fuscous with white on joints. Forewing: grayish fuscous ground color with yellow-buff overlaid on part of apical half except for dark fuscous borders to markings and large dark fuscous area mid-apically; dorsal margin with white crescent at midwing ending with small silver spot mesad; costal margin with 5 white bars each with silver spot mesad, with 4th and 5th from apex having longest white and silver combination and pointed oblique toward tornus; black spot on apex; silver spot at falcate indentation and along termen near tornus; tornus with small white spot with silver mesad bar; fringe fuscous, white distally except all white at falcate indentation; venter fuscous with dorsal white marks faintly repeated except distinct apical marks. **Hindwing:** fuscous; fringe fuscous; venter pale fuscous. **Abdomen:** fuscous with silvery scales on posterior of each segment; venter mostly white; 8th abdominal segment (Fig. 131) modified in male as genitalia hood with venter split, strongly sclerotized and with posterior point and projected central point each side, with numerous setae dorso- and latero-posteriorly from inside margin; coremata absent. **Genitalia** (Fig. 193): tuba analis long, narrow; tegumen elongate, stout; vinculum narrow, shallow U-shaped; saccus very long (subequal to distance from tuba analis to vinculum), narrow; valva elongate, setaceous, unevenly thickened, with small ventral projection from base of sacculus (Fig. 192); valva fused to anellus, with dorsal base fused as transtilla projected anteriorly as one central fin-shaped appendage; transtilla fused to strongly sclerotized, elongated, tubular anellus with pointed distal end; aedeagus (Fig. 194) very long (1.25 times saccus length), narrow, tapering to more slender apical point, straight; phallobase absent; cornutus absent (possibly long spines and deciduous); vesica without spicules; ductus ejaculatorius with hood distant from aedeagus (5 preparations examined). ·

Female (Fig. 119).- 3.8-4.2 mm. forewing length. Similar to male. **Genitalia** (Fig. 262): ovipositor short; papilla analis small, setaceous; apophyses short, thin, posterior pair 1.25 length anterior pair; 8th sternite unmodified; ostium bursae (Fig. 261) a small opening on a stout sclerotized projection surrounding posterior part of ductus bursae and extended near very convex and pointed 7th sternite; ductus bursae thin, posterior 1/3 sclerotized, then anterior 2/3 membranous; ductus seminalis emergent from base of ductus bursae as it merges into bursa; corpus bursae elongate-ovate, tapering, of moderate size, with minute spicules; signum (Fig. 260) as two small opposed patches of fused teeth-like spines directed anteriorly (3 preparations examined).

Types.- Holotype male: Lake Panasoffkee, Sumter Co., Florida, 11 May 1974, emerged 14 May 1974, K. W. Knopf (USNM). Paratypes (4 males, 4 females): **Florida.-** Alachua Co.: Archer Road Lab, 3 mi. [=4.8 km.] SW. Gainesville, 2 Apr 1976 (2 males, 1 female), 8 Apr 1975 (1 male), 5 May 1976 (1 male), 3 Aug 1975 (1 male), at blacklight, J. B. Heppner (JBH). Sumter Co.: Lake Panasoffkee, 11 May 1974 (2 females), emerged 15 & 17 May, K. W. Knopf (JBH). (Paratypes to BMNH and FSCA).

Distribution (Fig. 19).- Known only from Florida.

Flight period.- April to May; August.

Hosts.- Unknown. (Reared specimens emerged from a mixed sample of aquatic weeds incidental to an aquatic weed project of K. W. Knopf).

Biology.- Unknown.

Remarks.- Available specimens of **Diploschizia kimballi** are relatively uniform and have only minor variations of forewing markings in regard to size. As noted under **D. impigritella**, the two species are virtually identical in maculation and Florida specimens should be dissected to verify their identification. In viewing series of each species side by side it appears that **D. kimballi** has a slightly more slender forewing crescent mark and a darker apical quarter than do **D. impigritella** specimens.

The 8th abdominal segment is the most highly modified in the genus thus far known, yet lacks the coremata found in **D. impigritella**. The females that have been dissected all show what appear to be deciduous cornuti in their bursae but thus far no male has been found with such cornuti intact.

ILLUSTRATIONS

Explanation of Figures

The figures of morphological details of adult moths of Nearctic genera (Figs. 20–51) are generally similarly magnified on each plate; all have either a line scale on the figure or a magnification indicated in the caption. The wing venation figures (Figs. 52–57) are all drawn approximately the same size but have a scale line by each figure. The adult figures (Figs. 58–127), adult structures (Figs. 128–132), male genitalia (Figs. 133–202), and female genitalia (Figs. 203–268), all are enlarged to a similar size on each plate and actual dimensions are not indicated. Aedeagi often are shown at a higher magnification than the valvae but lengths are noted in the text for each species. Wing lengths are noted in the text for each species.

For each caption exclusive of all the figures up to Fig. 57, the locality data of the specimen used for the illustration is noted, together with the collection to which it belongs (in the case of holotypes, the collection it will be deposited in is the one noted). The slide number is noted where appropriate. Figures 20–56 refer to generic characters and only slide numbers are given if needed. Only specimens both photographed for the adult wing maculation and genitalia have the corresponding figure number cross referenced.

The species are usually figured in the sequence they are discussed in the text and checklist with the exception of several new species added to the revision after the initial completion of the plates: the additional figures are grouped into plates and inserted nearest their phylogenetic placement in the text.

The larval and pupal figures (Figs. 269–298) are based on single specimens. In the chaetotaxic maps of larval segments and heads, the nomenclature and abbreviations of Hinton (1946) are used.

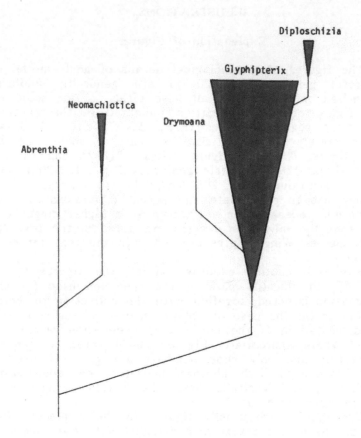

Fig. 1. Phylogeny of Nearctic Glyphipterigidae.

2

Fig. 2. Distribution map of *Abrenthia cuprea* Busck.

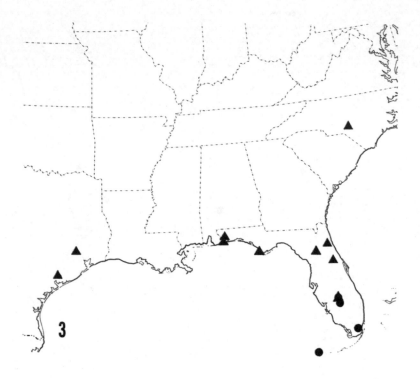

Fig. 3. Distribution map of *Neomachlotica spiraea* Heppner (●) and *Drymoana blanchardi* Heppner (▲).

Fig. 4. Distribution map of *Glyphipterix brauni* Heppner (o), *Glyphipterix circumscriptella circumscriptella* Chambers (•), and *Glyphipterix circumscriptella apacheana* Heppner (▲).

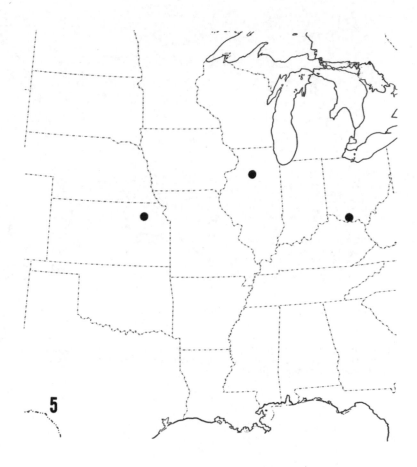

Fig. 5. Distribution map of *Glyphipterix quadragintapunctata* Dyar.

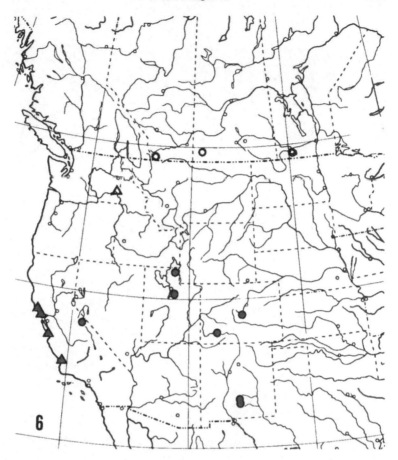

Fig. 6. Distribution map of *Glyphipterix powelli powelli* Heppner (▲), *Glyphip-terix powelli jucunda* Heppner (△), *Glyphipterix urticae urticae* Heppner (●), and *Glyphipterix urticae sylviborealis* Heppner (o).

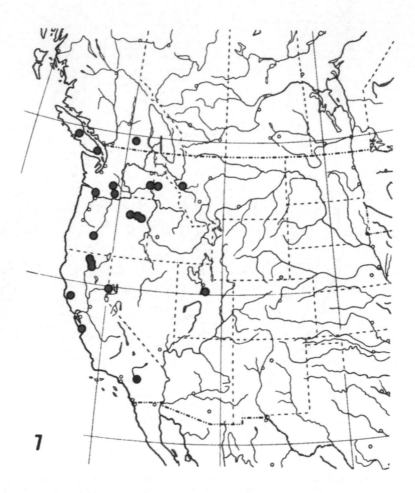

7

Fig. 7. Distribution map of *Glyphipterix bifasciata* Walsingham.

Fig. 8. Distribution map of *Glyphipterix hypenantia* Heppner (▲), *Glyphipterix yosemitella* Heppner (o), and *Glyphipterix unifasciata* Walsingham (•).

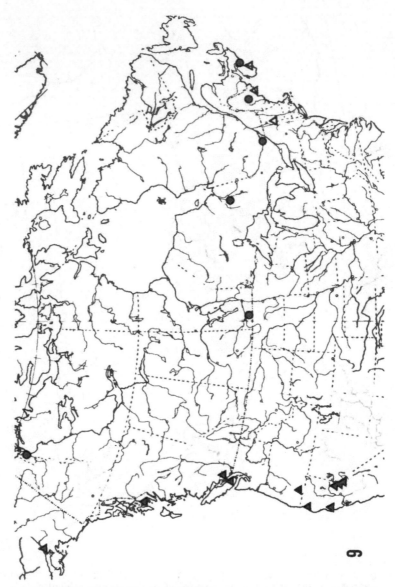

Fig. 9. Distribution map of *Glyphipterix haworthana* (Stephens) (•), *Glyphipterix sistes sistes* Heppner (▲), and *Glyphipterix sistes viridimontis* Heppner (△).

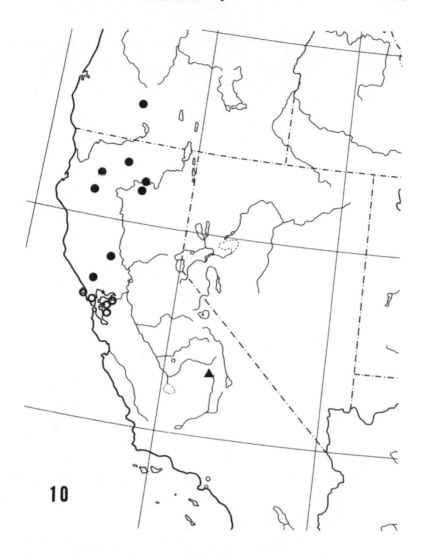

10

Fig. 10. Distribution of *Glyphipterix californiae* Walsingham (•), *Glyphipterix feniseca* Heppner (o), and *Glyphipterix sierranevadae* Heppner (▲).

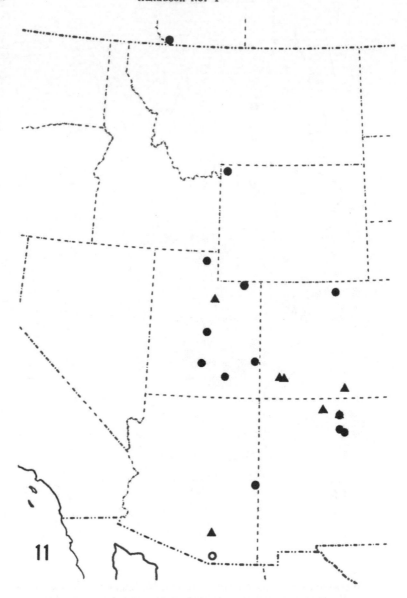

Fig. 11. Distribution of *Glyphipterix juncivora* Heppner (•), *Glyphipterix arizonensis* Heppner (o), and *Glyphipterix roenastes* Heppner (▲).

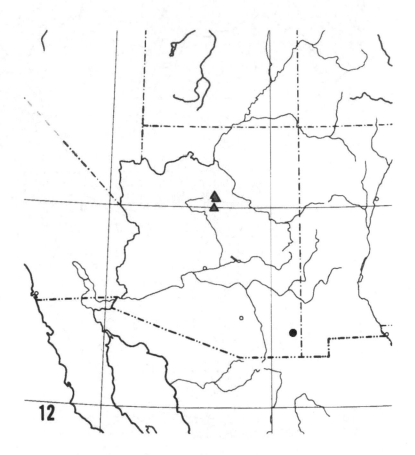

Fig. 12. Distribution map of *Glyphipterix chiricahuae* Heppner (●), and *Glyphip-terix hodgesi* Heppner (▲).

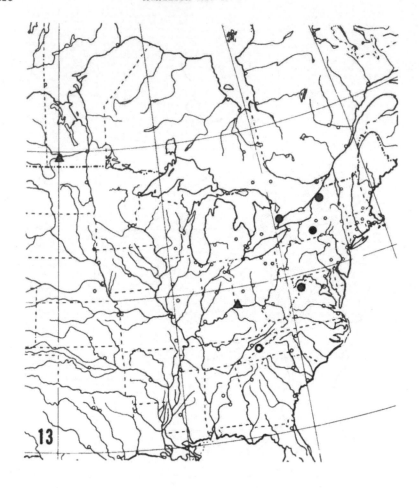

Fig. 13. Distribution map of *Glyphipterix saurodonta* Meyrick (●), *Glyphipterix cherokee* Heppner (o), and *Glyphipterix chambersi* Heppner (▲).

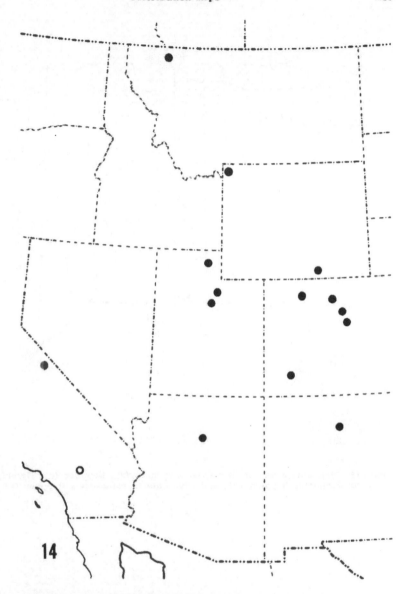

Fig. 14. Distribution map of *Glyphipterix montisella* Chambers (•) and *Glyphip-terix flavimaculata* Heppner (o).

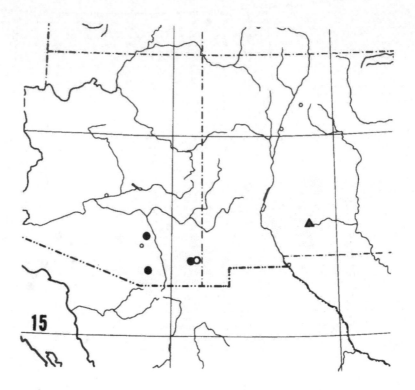

Fig. 15. Distribution map of *Glyphipterix melanoscirta* Heppner (o), *Glyphip-*
terix santaritae Heppner (o), and *Glyphipterix ruidosensis* Heppner (▲).

16

Fig. 16. Distribution map of *Diploschizia lanista* (Meyrick).

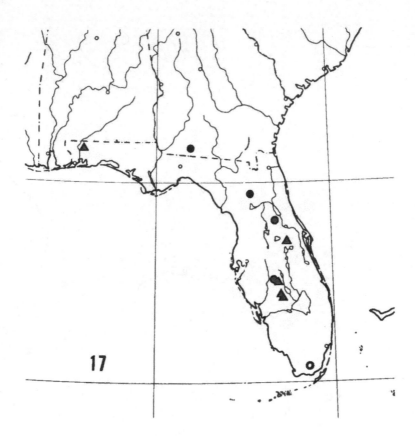

Fig. 17. Distribution map of **Diploschizia** minimella Heppner (▲), **Diploschizia**
habecki Heppner (●), and **Diploschizia regia** Heppner (o).

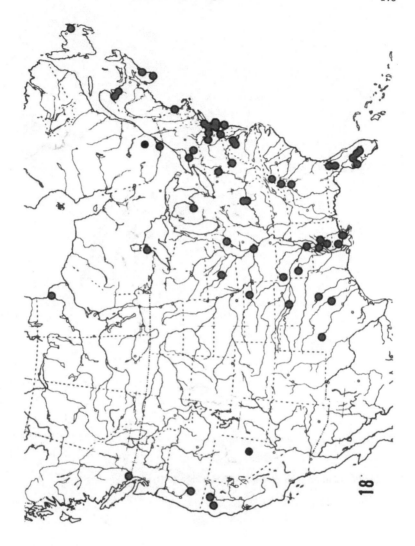

Fig. 18. Distribution map of *Diploschizia impigritella* (Clemens).

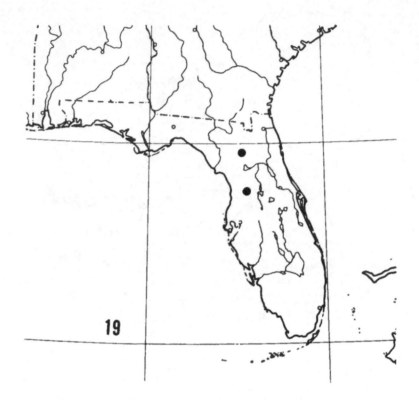

Fig. 19. Distribution map of **Diploschizia kimballi** Heppner.

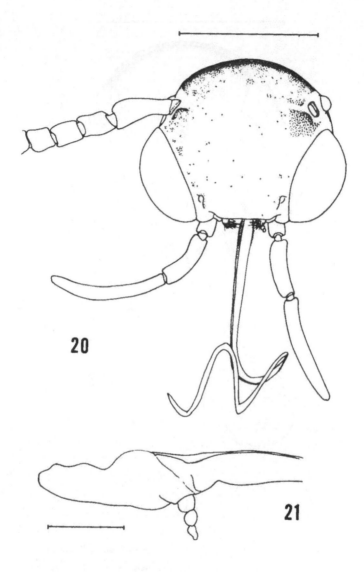

20

21

Figs. 20-21. Head morphology of *Abrenthia*. 20. *Abrenthia cuprea* Busck, male (scale line = 0.5 mm) (slide USNM 77720). 21. Details of left maxilla (scale line = 0.1 mm).

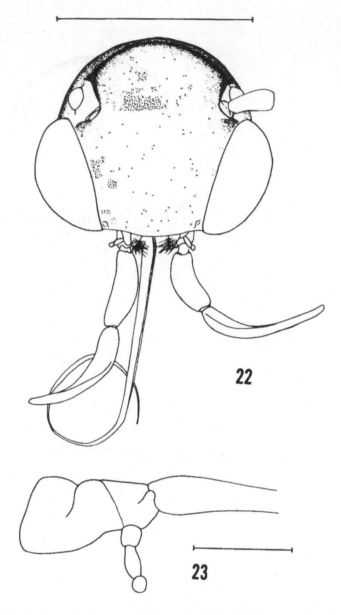

Figs. 22-23. Head morophology of *Neomachlotica*. 22. *Neomachlotica spiraea* Heppner, female (scale line = 0.5 mm) (slide USNM 77721). 23. Detail of left maxilla (scale line = 0.1 mm).

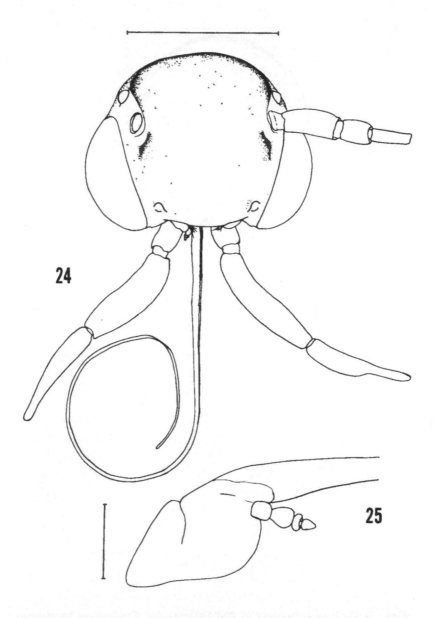

Figs. 24-25. Head morphology of *Drymoana*. 24. *Drymoana blanchardi* Heppner, female (scale line = 0.5 mm) (slide USNM 77722). 25. Detail of left maxilla (scale line = 0.1 mm).

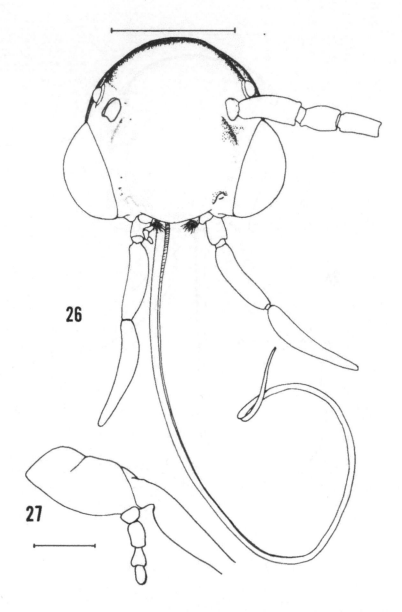

Figs. 26-27. Head morphology of *Glyphipterix*. 26. *Glyphipterix bergstraes-serella* (Fabricius), male (scale line = 0.5 mm) (slide USNM 77724). 27. Detail of left maxilla (scale line = 0.1 mm).

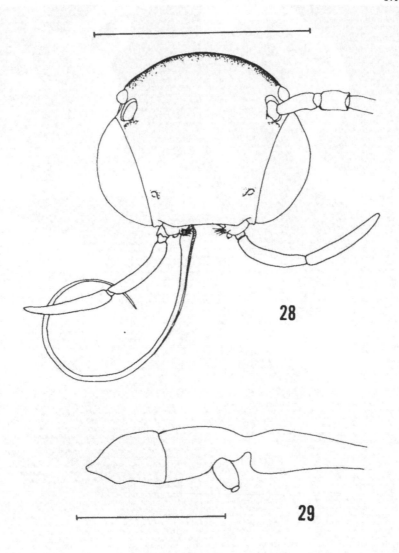

28

29

Figs. 28–29. Head morphology of *Diploschizia*. 28. *Diploschizia impigritella* (Clemens), male (scale line = 0.5 mm) (slide USNM 77825). 29. Details of left maxilla (scale line = 0.1 mm).

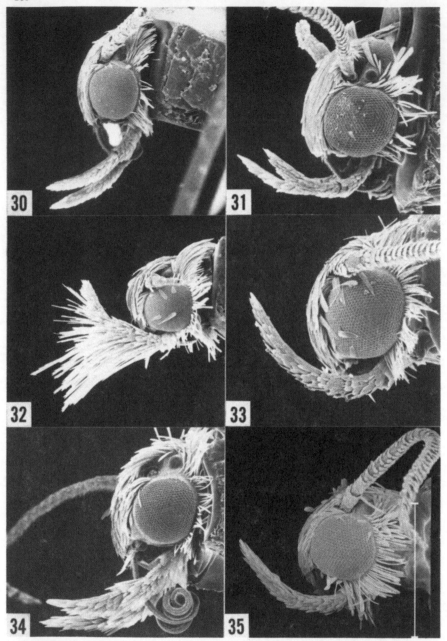

Figs. 30-35. Head profiles. 30. *Abrenthia cuprea* Busck, male (62 X) (slide USNM 77341). 31. *Neomachlotica spiraea* Heppner, male (100 X) (slide USNM 77342). 32. *Drymoana blanchardi* Heppner, female (55 X) (slide USNM 77338). 33. *Glyphipterix quadragintapunctata* Dyar, male (80 X) (slide USNM 77339). 34. *Glyphipterix bergstraesserella* (Fabricius), male (76 X) (slide USNM 77343). 35. *Diploschizia impigritella* (Clemens), male (110 X) (scale = 0.5 mm) (slide USNM 77340).

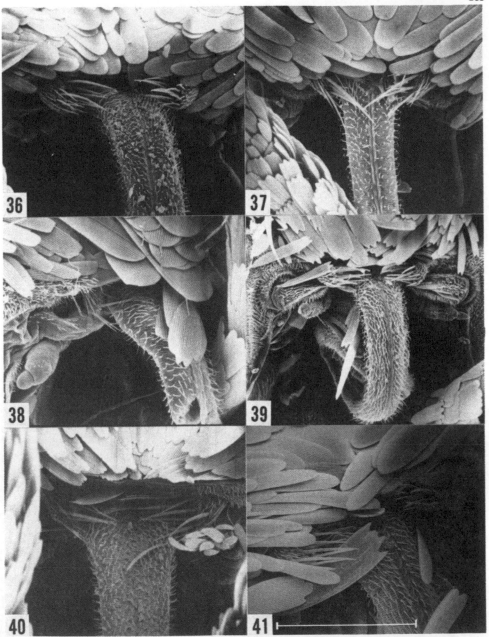

Figs. 36–41. Haustellata. 36. *Abrenthia cuprea* Busck, male (325 X) (slide USNM 77341). 37, *Neomachlotica spiraea* Heppner, male (420 X) (slide USNM 77342). 38. *Drymoana blanchardi* Heppner, female (415 X) (slide USNM 77338). 39. *Glyphipterix quadragintapunctata* Dyar, male (235 X) (slide USNM 77339). 40. *Glyphipterix bergstraesserella* (Fabricius), male (420 X) (slide USNM 77343). 41. *Diploschizia impigritella* (Clemens), male (scale = 0.1 mm) (595 X) (slide USNM 77340).

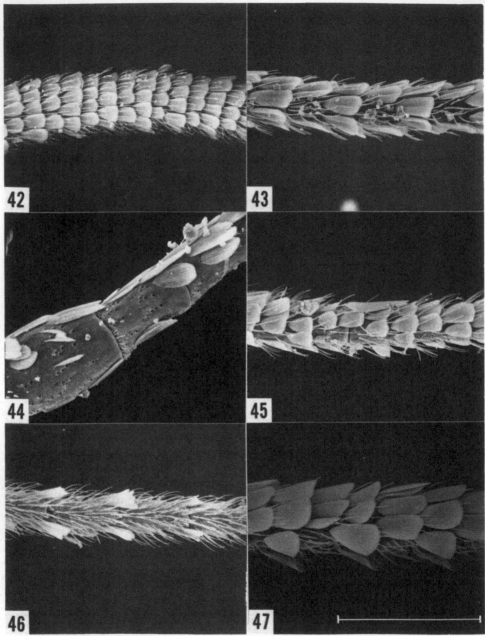

Figs. 42-47. Antennae. 42. *Abrenthia cuprea* Busck, male (200 X) (slide USNM 77341). 43. *Neomachlotica spiraea* Heppner, male (400 X) (slide USNM 77342). 44. *Drymoana blanchardi* Heppner, female (460 X) (slide USNM 77338). 45. *Glyphipterix quadragintapunctata* Dyar, male (265 X) (slide USNM 77339). 46. *Glyphipterix bergstraesserella* (Fabricius), male (280 X) (slide USNM 77343). 47. *Diploschizia impigritella* (Clemens), male (scale = 0.1 mm) (605 X) (slide USNM 77340).

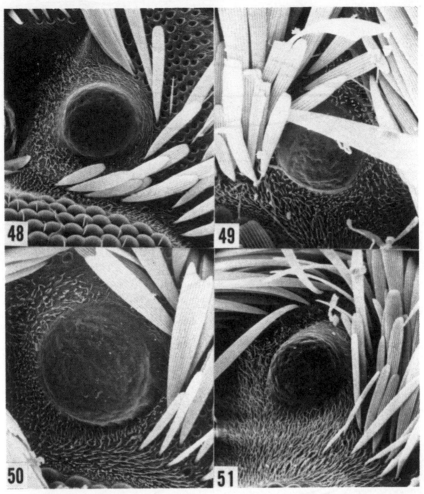

Figs. 48-51. Ocelli. 48. *Neomachlotica spiraea* Heppner, male (650 X) (slide USNM 77342). 49. *Drymoana blanchardi* Heppner, female (540 X) (slide USNM 77338). 50. *Glyphipterix quadragintapunctata* Dyar, male (620 X) (slide USNM 77339). 51. *Glyphipterix bergstraesserella* (Fabricius), male (475 X) (slide USNM 77343).

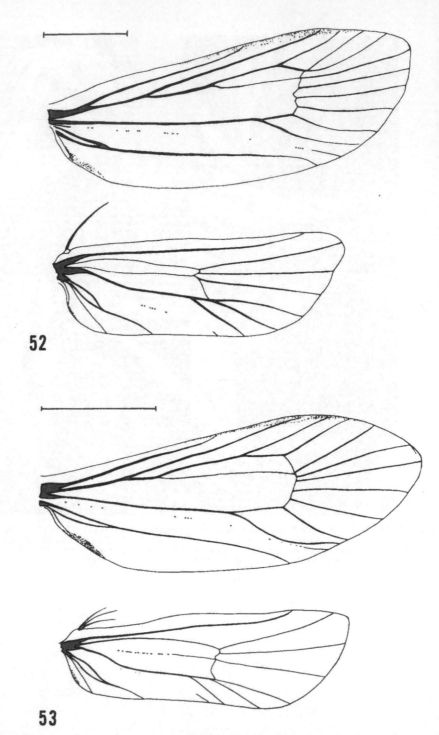

52

53

Figs. 52-53. Wing venation of genera. 52. *Abrenthia cuprea* Busck, (scale line = 1 mm) (slide USNM 77021). 53. *Neomachlotica spiræa* Heppner, (scale line = 1 mm) (slide USNM 77225).

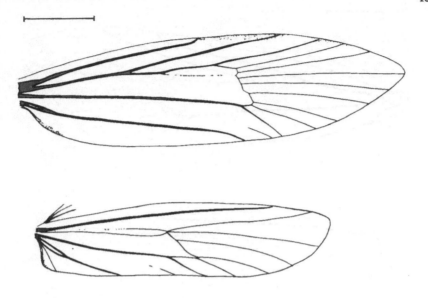

54

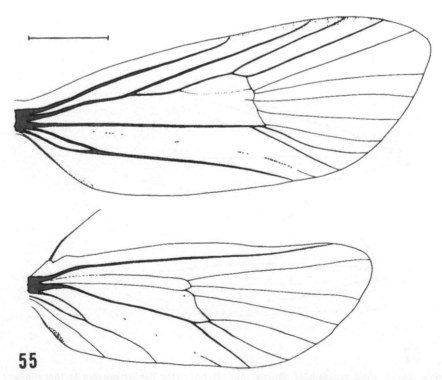

55

Figs. 54-55. Wing venation of genera. 54. *Drymoana blanchardi* Heppner, (scale line = 1 mm) (slide USNM 77226). 55. *Glyphipterix quadragintapunctata* Dyar, (scale line = 1 mm) (slide USNM 77227).

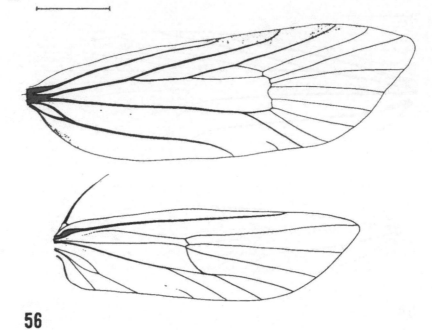

56

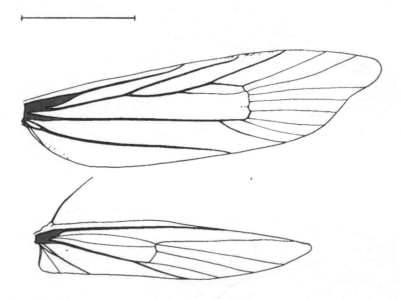

57

Figs. 56-57. Wing venation of genera. 56. *Glyphipterix bergstraesserella* (Fabricius),
(scale line = 1 mm). 57. *Diploschizia impigritella* (Clemens), (scale line = 1 mm)
(slide USNM 77467).

Figs. 58–65. Adults. 58. **Abrenthia cuprea** Busck, male. Cincinnati, [Hamilton Co.], Ohio, 10 Jun 1927, A. F. Braun. (ANSP). 59. **Abrenthia cuprea** Busck, female, Plummers Is., [Montgomery Co.], Maryland, 8 Jul 1968, P. J. Spangler. (USNM) (see figs. 203–204, female genitalia). 60. **Neomachlotica spiraea** Heppner, male (holotype), Fisheating Cr., 2 mi. SE. Palmdale, Glades Co., Florida, 6 May 1975, J. B. Heppner, on **Pluchea purpurascens** flowers. (USNM) (see figs. 135–137, male genitalia). 61. **Neomachlotica spiraea** Heppner, female (paratype). Fisheating Cr., 2 mi. SE. Palmdale, Glades Co., Florida, 6 May 1975, J. B. Heppner, on **Pluchea purpurascens** flowers. (USNM) (see figs. 205–207, female genitalia) (right wings reversed). 62. **Drymoana blanchardi** Heppner, male (holotype). Deutschburg, Jackson Co., Texas, 7 Oct 1974, A. & M. E. Blanchard. (USNM) (see fig. 138, male genitalia) (right wings reversed). 63. **Drymoana blanchardi** Heppner, female (paratype). Camp Strake, Montgomery Co., Texas, 9 Sep 1975, A. & M. E. Blanchard. (USNM) (see figs. 208–209, female genitalia). 64. **Glyphipterix quadragintapunctata** Dyar, male. Beaver Pond, Adams Co., Ohio, 11 Jun 1930, A. F. Braun. (ANSP) (right wings reversed). 65. **Glyphipterix quadragintapunctata** Dyar, female. Mineral Springs, Adams Co., Ohio, 27 Jun 1931, A. F. Braun. (ANSP) (right wings reversed).

Figs. 66–73. Adults. 66. **Glyphipterix powelli powelli** Heppner, male (paratype). Inverness, Marin Co., California, 15 Mar 1959, D. Burdick. (CAS) (see figs. 142–143, male genitalia) (right wings reversed). 67. **Glyphipterix powelli powelli** Heppner, female (paratype). Inverness, Marin Co., California, 5 Apr 1959, D. Burdick. (UCB) (see figs. 212–213, female genitalia). 68. **Glyphipterix powelli jucunda** Heppner, male (holotype). Pullman, [Whitman Co.], Washington, 26 May 1951, R. B. Spurrier. (WSU) (see figs. 144–145, male genitalia) (right wings reversed). 69. **Glyphipterix powelli jucunda** Heppner, female (paratype). Pullman, [Whitman Co.], Washington, 2 Jun 1965, R. D. Akre. (WSU) (see figs. 214–215, female genitalia). 70. **Glyphipterix urticae urticae** Heppner, male (holotype). South Platte R., 1½ mi. SW. Lake George, Park Co., Colorado, 1 Jul 1976, J. B. Heppner, on **Urtica** sp. (USNM). 71. **Glyphipterix urticae urticae** Heppner, female (paratype). South Platte R., 1½ mi. SW. Lake George, Park Co., Colorado, 1 Jul 1976, J. B. Heppner, on **Urticae** sp. (USNM) (right wings reversed). 72. **Glyphipterix urticae sylviborealis** Heppner, male (holotype). Waterton Lakes, Alberta, Canada, 28 Jun 1923, J. H. McDunnough. (CNC). 73. **Glyphipterix urticae sylviborealis** Heppner, female (paratype). Waterton Lakes, Alberta, Canada, 30 Jun 1923, J. H. McDunnough. (CNC).

Figs. 74–81. Adults. 74. **Glyphipterix bifasciata** Walsingham, male. 5 mi. E. McCloud, Siskiyou Co., California, 14 Jul 1962, J. A. Powell. (UCB) (image reduced). 75. **Glyphipterix bifasciata** Walsingham, female. 2.5 mi. W. Ft. Simcoe, Yakima Co., Washington, 31 Jul 1962, J. F. G. Clarke. (USNM) (right wings reversed). 76. **Glyphipterix unifasciata** Walsingham, male. Novato, Marin Co., California, 6 May 1962, D. C. Rentz. (UCB) (right wings reversed). 77. **Glyphipterix unifasciata** Walsingham, female. Novato, Marin Co., California, 6 May 1962, D. C. Rentz. (UCB) (see figs. 224–225, female genitalia). 78. **Glyphipterix hypenantia** Heppner, male (paratype). 1 mi. SE. Bartle, Siskiyou Co., California, 11–14 Jun 1974, J. A. Powell. (UCB) (right wings reversed). 79. **Glyphipterix circumscriptella apacheana** Heppner, male (holotype). Stewart Camp, 1 mi. S. Portal, [Chiricahua Mts.], Cochise Co., Arizona, 26–29 Aug 1971, J. T. Doyen. (CAS). 80. **Glyphipterix circimscriptella circumscriptella** Chambers, male. Putnam Co., Illinois, 21 Jun 1949, M. O. Glenn. (USNM) (right wings reversed) 81. **Glyphipterix circumscriptella circumscriptella** Chambers, female. Essex Co. Park, New Jersey, 1 Jul, W. D. Kearfott. (USNM) (see figs. 226–277, female genitalia).

190

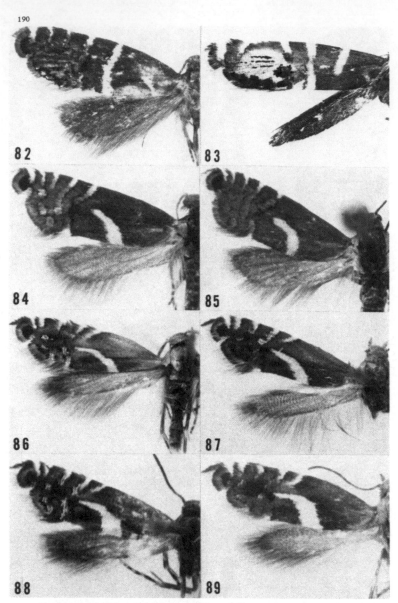

Figs. 82–89. Adults. 82. **Glyphipterix yosemitella** Heppner, male (holotype). 4 mi. S. Mather, Tuolumne Co., California, 12 Jun 1961, J. A. Powell. (CAS) (see figs. 160–161, male genitalia). 83. **Glyphipterix brauni** Heppner, male (holotype). Grant Teton Natl. Pk., [Teton Co.], Wyoming, 6 Jul 1959, A. F. Braun. (ANSP) (see figs. 158–159, male genitalia). 84. **Glyphipterix haworthana** (Stephens), male. Mer Bleue, Ontario, Canada, 13 May 1932, W. J. Brown. (USNM). 85. **Glyphipterix haworthana** (Stephens), female. Big Indian Lake, Halifax, Nova Scotia, Canada, 25 May 1959, D. C. Ferguson. (NSM). 86. **Glyphipterix sistes sistes** Heppner, male (paratype). Duncan, Vancouver Is., British Columbia, Canada, 11 Aug 1925, A. W. Hanham. (USNM). 87. **Glyphipterix sistes sistes** Heppner, female (paratype). Duncan, Vancouver Is., British Columbia, Canada, 31 Jul 1925, A. W. Hanham. (USNM) (see figs. 252–253, female genitalia. 88. **Glyphipterix sistes viridimontis** Heppner, male (paratype). Chittenden Brook Cpgd., Green Mts., Rutland Co., Vermont, 18 Aug 1973, on **Solidago** flowers, J. B. Heppner. (USNM) (right wings reversed). 89. **Glyphipterix sistes viridimontis** Heppner, female (paratype). Chittenden Brook Cpgd., Green Mts., Rutland Co., Vermont, 18 Aug 1973, on **Solidago** flowers, J. B. Heppner. (USNM).

Figs. 90–95. Adults. 90. **Glyphipterix roenastes** Heppner, male (holotype). Silverton, [San Juan Co.], Colorado, 16–23 Jul, [9200']. (USNM) (see figs. 173–174, male genitalia). 91. **Glyphipterix roenastes** Heppner, female (paratype). 15 mi. SE. Heber City, Wasatch Co., Utah, 7 Jul 1976, J. B. Heppner. (USNM) (see figs. 236–237, female genitalia). 92. **Glyphipterix californiae** Walsingham, male. Big Flat Cpgd., Trinity Co., California, 29 Jun 1974, E. Rogers. (UCB). 93. **Glyphipterix californiae** Walsingham, female. Mt. Shasta City, Siskiyou Co., California, 1 Jul 1970, P. Rude. (UCB) (see figs. 238–239, female genitalia). 94. **Glyphipterix juncivora** Heppner, male (paratype). Bowery Creek Cpgd., Fish Lake, Sevier Co., Utah, 11 Jul 1976, 9000', on Juncus, J. B. Heppner. (USNM). 95. **Glyphipterix juncivora** Heppner, female (paratype). Lake Hill Cpgd., 7 mi. E. Ephraim, Sanpete Co., Utah, 10 Jul 1976, 8500', on Juncus, J. B. Heppner (USNM) (right wings reversed).

Figs. 96–101. Adults. 96. **Glyphipterix feniseca** Heppner, male (holotype). North Beach, Pt. Reyes Natl. Seashore, Marin Co., California, 11 May 1974, G. Bunker. (CAS) (see figs. 167–168, male (genitalia). 97. **Glyphipterix feniseca** Heppner, female (paratype). Alpine Lake, Marin Co., California, 25 Apr 1958, J. A. Powell. (UCB) (see figs. 230–231, female genitalia). 98. **Glyphipterix sierranevadae** Heppner, male (holotype). Mineral King, Tulare Co., California, 1–7 Jul. (USNM) (see figs. 169–170, male genitalia). 99. **Glyphipterix sierranevadae** Heppner, female (paratype). Mineral King, Tulare Co., California, 24–31 Jul. (USNM) (see figs. 232–233, female genitalia). 100. **Glyphipterix arizonensis** Heppner, male (holotype). Madera Cyn., Santa Rita Mts., Santa Cruz Co., Arizona, 27–30 Jul 1947, L. M. Martin. (LACM) (see figs. 171–172, male genitalia). 101. **Glyphipterix arizonensis** Heppner, female (paratype). "Arizona", 1882, Morrison, Walsingham Coll. 35562. (BMNH) (see figs. 234–235, female genitalia).

Figs. 102–107. Adults. 102. **Glyphipterix chiricahuae** Heppner, male (holotype), S. W. R. S. [= Southwest Res. Sta.], 5 mi. W. Portal, [Chiricahua Mts.], Cochise Co., Arizona, 9 Nov 1964, 5400', V. D. Roth. (AMNH) (see figs. 177–178, male genitalia). 103. **Glyphipterix chiricahuae** Heppner, female (paratype). S. W. R. S. (USNM) (right wings reversed). 104. **Glyphipterix saurodonta** Meyrick, female. Ottawa, Ontario, Canada, 17 Aug 1906, C. H. Young. (CNC). 105. **Glyphipterix cherokee** Heppner, female (holotype). Great Smoky Mts., Tennessee, 23 Aug 1950, 6000', G. S. Walley. (CNC) (see figs. 254–255, female genitalia). 106. **Glyphipterix chambersi** Heppner, female (holotype). Marmont, R[oun]thwaite, Manitoba, Canada, 8 Aug 1905. (USNM). 107. **Glyphipterix flavimaculata** Heppner, male (holotype). Hathaway Cr., San Bernardino Mts., [San Bernardino Co.], California, 26 Jul 1942, 8000'. (USNM) (see figs. 162–163, male genitalia).

194

Figs. 108–113. Adults. 108. **Glyphipterix hodgesi** Heppner, male (holotype). Hart Prairie, 10 mi. NW.
Flagstaff, [San Francisco Mts.], Coconino Co., Arizona, 23 Aug 1961, 8500', R. W. Hodges. (USNM)
(right wings reversed). 109. **Glyphipterix hodgesi** Heppner, female (paratype). Hart Prairie, 10 mi.
NW. Flagstaff, [San Francisco Mts.], Coconino Co., Arizona, 28 Aug 1961, 8500', R. W. Hodges.
(USNM) (see figs. 242–243, female genitalia) (right wings reversed). 110. **Glyphipterix montisella**
Chambers, male. Hidden Valley, Rocky Mtn. Natl. Pk., Colorado, 11 Aug 1929, A. F. Braun. (USNM).
111. **Glyphipterix montisella** Chambers, female. Hidden Valley, Rocky Mtn. Natl. Pk., Colorado, 11
Aug 1929, A. F. Braun. (USNM). 112. **Glyphipterix santaritae** Heppner, male (holotype). Madera
Cyn., Santa Rita Mts., Santa Cruz Co., Arizona, 1–3 Aug 1970, P. Rude. (CAS) (right wings reversed)
113. **Glyphipterix santaritae** Heppner, female (paratype). Bear Cyn., Santa Catalina Mts., [Pima
Co.], Arizona, 2 Aug 1970, J. A. Powell. (UCB) (see figs. 246–247, female genitalia).

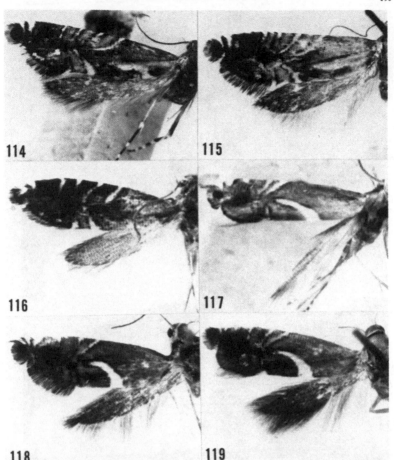

Figs. 114–119. Adults. 114. **Glyphipterix ruidosensis** Heppner, male (holotype). Ruidoso Cyn., [Sacramento Mts., Otero–Lincoln Co.], New Mexico, 30 Sep 1916, C. Heinrich. (USNM) (see fig. 189, male genitalia). 115. **Glyphipterix ruidosensis** Heppner, female (paratype). Ruidoso Cyn., [Sacramento Mts., Otero–Lincoln Co.], New Mexico, 4 Oct 1916, C. Heinrich. (USNM) (see figs. 258–259, female genitalia). 116. **Glyphipterix melanoscirta** Heppner, male (holotype). S. W. R. S. [= Southwestern Res. Sta.], 5 mi. W. Portal, [Chiricahua Mts.], Cochise Co., Arizona, 21 Jul 1967, 5400'. (AMNH) (see fig. 164, male genitalia). 117. **Diploschizia regia** Heppner, male (holotype). Royal Palm State Pk., [= Royal Palm Hammock, Everglades Natl. Pk., Dade Co.], Florida, Jan 1930, F. M. Jones. (USNM) (see figs. 190–191, male genitalia). 118. **Diploschizia kimballi** Heppner, male (holotype). Lake Panasoffkee, Sumter Co., Florida, 11 May 1974, emerged 14 May 1974, K. W. Knopf. (USNM) (see figs. 193–194, male genitalia) (right wings reversed). 119. **Diploschizia kimballi** Heppner, female (paratype). Lake Panasoffkee, Sumter Co., Florida, 11 May 1974, emerged 17 May 1974, K. W. Knopf. (USNM).

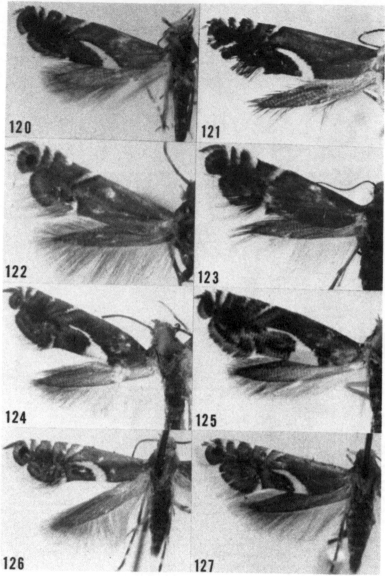

Figs. 120–127. Adults. 120. **Diploschizia lanista** (Meyrick), male. Southern Pines, [Moore Co.], North North Carolina. (USNM). 121. **Diploschizia lanista** (Meyrick), female. Pensacola, [Escambia Co.], Florida, 30 Sep 1961, S. O. Hills. (FSCA). 122. **Diploschizia minimella** Heppner, male (holotype). Archbold Biological Sta., [10 mi.S.] Lake Placid, Highlands Co., Florida, 16–22 May 1964, R. W. Hodges. (USNM) (see figs. 197–198, male genitalia) (right wings reversed). 123. **Diploschizia mini-mella** Heppner, female (paratype). Archbold Biological Sta., [10 mi. S.] Lake Placid, Highlands Co., Florida, 29 Mar 1959, R. W. Hodges. (USNM). 124. **Diploschizia habecki** Heppner, male (paratype). Highlands Hammock St. Pk., Highlands Co., Florida, reared ex **Rhynchospora corniculata**, 26 May 1974, J. B. Heppner. (USNM). 125. **Diploschizia habecki** Heppner, female (paratype). Bivens Arm Lake, 3 mi. SW. Gainesville, Alachua Co., Florida, reared ex **Rhynchospora corniculata**, 17 Jul 1974, J. B. Heppner. (USNM). 126. **Diploschizia impigritella** (Clemens), male. Cincinnati, [Hamilton Co.], Ohio, 30 Jul 1907, A. F. Braun. (ANSP) (right wings reversed). 127. **Diploschizia impigritella** (Clemens), female. Cincinnati, [Hamilton Co.], Ohio, 14 Aug 1907, A. F. Braun. (ANSP) (right wings reversed).

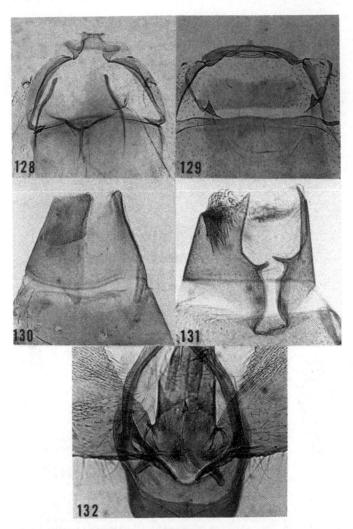

Figs. 128–132. Adult Structures. 128. **Glyphipterix californiae** Walsingham, male, ventral abdominal articulation (tineoid type). (slide JBH 528) (ANSP). 129 **Prochoreutis inflatella** (Clemens), male, ventral abominal articulation (tortricoid type) in Choreutidae. (slide USNM 77118) (USNM). 130. **Diploschizia impigritella** (Clemens), male, 8th abdominal sternite of male, ventral view. (slide JBH 303) (ANSP). 131. **Diploschizia kimballi** Heppner, male, 8th abdominal sternite of male, ventral view. (slide USNM 77822) (USNM). 132. **Glyphipterix californiae** Walsingham, male, tubular anellus-valval articulation. (slide JBH 528) (ANSP).

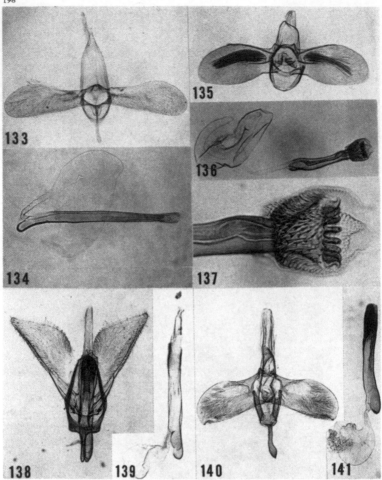

Figs. 133–141. Male genitalia. 133. **Abrenthia cuprea** Busck, male (lectotype). Roxboro, [Philadelphia Co.], Pennsylvania, 21 Jun, F. Haimbach. (USNM) (slide USNM 77247). 134. **Abrenthia cuprea** Busck, male (lectotype) aedeagus, same data. 135. **Neomachlotica spiraea** Heppner, male (holotype). Fish-eating Cr., 2 mi. SE. Palmdale, Glades Co., Florida, 6 May 1975, on **Pluchea purpurascens** flowers, J. B. Heppner. (USNM) (slide USNM 77151) (see fig. 60, adult). 136. **Neomachlotica spiraea** Heppner male (holotype) aedeagus, same data. 137. **Neomachlotica spiraea** Heppner, male (holotype) aedeagus tip, same data. 138. **Drymoana blanchardi** Heppner, male (holotype). Deutschburg, Jackson Co., Texas, 7 Oct 1974, A. & M.E. Blanchard. (USNM) (slide USNM 77149) (see fig. 62, adult). Aedeagus in situ. 139. **Drymoana blanchardi** Heppner, male (paratype) aedeagus. Southern Pines, [Moore Co.], North Carolina, 8–15 Sep. (USNM) (slide USNM 77144). 140. **Glyphipterix quadragintapunctata** Dyar, male. Onaca, [Pottawatomie Co.], Kansas. (USNM) (slide USNM 77130). 141. **Glyphipterix quadragintapunctata** Dyar, male aedeagus, same data.

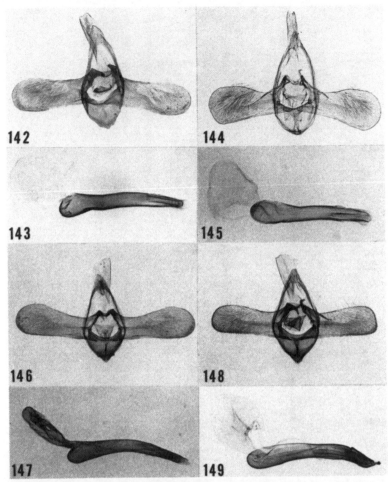

Figs. 142-149. Male genitalia. 142. **Glyphipterix powelli powelli** Heppner, male (paratype). Inverness, Marin Co., California, 15 Mar 1959, D. Burdick. (CAS) (slide JBH 380) (see fig. 66, adult). 143. **Glyphipterix powelli powelli** Heppner, male (holotype) aedeagus, same data. 144. **Glyphipterix powelli jucunda** Heppner, male (holotype). Pullman, [Whitman Co.], Washington, 26 May 1951, R. B. Spurrier (WSU) (slide JBH 522) (see fig. 68, adult). 145. **Glyphipterix powelli jucunda** Heppner, male (holotype) aedeagus, same data. 146. **Glyphipterix urticae urticae** Heppner, male (paratype), S. Fork Bonito Cr. Cpgd., Sacramento Mts., Lincoln Co., New Mexico, 6 Jul 1977, on **Urtica** sp., J. B. Heppner. (USNM). (slide USNM 77824). 147. **Glyphipterix urticae urticae** Heppner, male (paratype) aedeagus, same data. 148. **Glyphipterix urticae sylviborealis** Heppner, male (paratype). Cypress Hills, Maple Cr., Saskatchewan, Canada, 3 Jun 1926,C. H. Young. (CNC) (slide JBH 1354). 149. **Glyphipterix urticae sylviborealis** Heppner, male (paratype) aedeagus, same data.

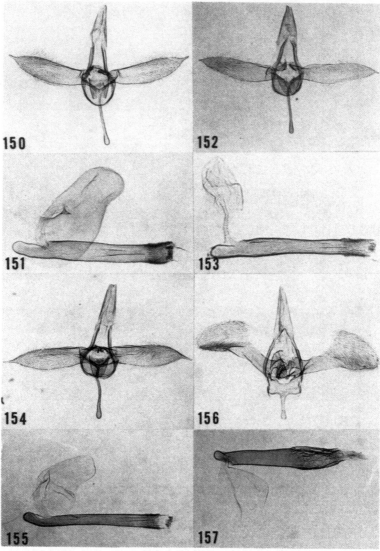

Figs. 150–157. Male genitalia. 150. **Glyphipterix bifasciata** Walsingham, male. Shasta Retreat, [Siskiyou Co.], California, 16–23 Jun. (USNM) (slide USNM 77123). 151. **Glyphipterix bifasciata** Walsingham, male aedeagus, same data. 152. **Glyphipterix hypenantia** Heppner, male (holotype). 1 mi. SE. Bartle, Siskiyou Co., California, 11–14 Jun 1974, J. A. Powell. (CAS) (slide JBH 466). 153. **Glyphipterix hypenantia** Heppner, male (holotype) aedeagus, same data. 154. **Glyphipterix unifasciata** Walsingham, male. Novato, Marin Co., California, 6 May 1962, D. C. Rentz. (CAS) (slide JBH 374). 155. **Glyphipterix unifasciata** Walsingham, male aedeagus, same data. 156. **Glyphipterix circumscriptella circumscriptella** Chambers, male. Essex Co. Park, New Jersey, 7 Jul, W. D. Kearfott. (USNM) (slide USNM 77135). 157. **Glyphipterix circumscriptella circumscriptella** Chambers, male aedeagus, same data.

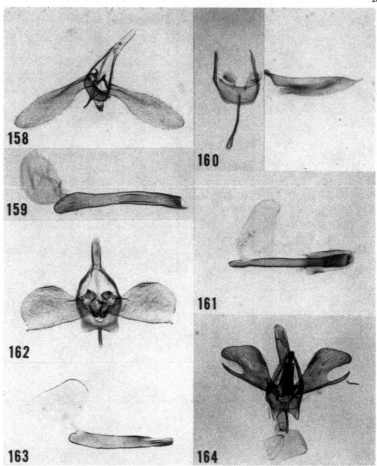

Figs. 158–164. Male genitalia. 158. **Glyphipterix brauni** Heppner, male (holotype). Grand Teton Natl. Pk., [Teton Co.], Wyoming, 6 Jul 1959, A. F. Braun. (ANSP) (slide JBH 1398) (see fig. 83, adult). 159. **Glyphipterix brauni** Heppner, male (holotype) aedeagus, same data. 160. **Glyphipterix yosemitella** Heppner, male (holotype). 4 mi. S. Mather, Tuolumne Co., California, 12 Jun 1961, J. A. Powell. (CAS) (slide JBH 464, damaged; valva aligned near true position) (see fig. 82, adult). 161. **Glyphipterix yosemitella** Heppner, male (holotype) aedeagus, same data. 162. **Glyphipterix flavimaculata** Heppner, male (holotype). Hathaway Cr., San Bernardino Mts., [San Bernardino Co.], California, 26 Jul 1942, 8000'. (USNM) (slide USNM 77143) (see fig. 107, adult). 163. **Glyphipterix flavimaculata** Heppner, male (holotype) aedeagus, same data. 164. **Glyphipterix melanoscirta** Heppner, male (holotype). S. W. R. S. [= Southwestern Res. Sta.], 5 mi. W. Portal, [Chiricahua Mts.], Cochise Co., Arizona, 21 Jul 1967, 5400'. (AMNH) (slide JBH 569) (see fig. 116, adult). Aedeagus in situ.

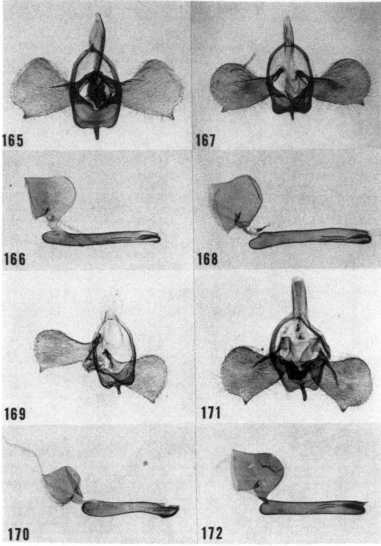

Figs. 165–172. Male genitalia. 165. **Glyphipterix juncivora** Heppner, male (holotype). Lake Hill Cpgd., 7 mi. E. Ephraim, Sanpete Co., Utah, 10 Jul 1976, 8500', on *Juncus*, J. B. Heppner. (USNM) (slide USNM 77790). 166. **Glyphipterix juncivora** Heppner, male (holotype) aedeagus, same data. 167. **Glyphipterix feniseca** Heppner, male (paratype). Alpine Lake, Marin Co., California, 25 Apr 1958, J. A. Powell. (UCB) (slide JBH 1401). 168. **Glyphipterix feniseca** Heppner, male (paratype) aedeagus, same data. 169. **Glyphipterix sierranevadae** Heppner, male (holotype). Mineral King, Tulare Co., California, 1–7 Jul. (USNM) (slide USNM 77795) (see fig. 98, adult). 170 **Glyphipterix sierranevadae** Heppner, male (holotype) aedeagus, same data. 171. **Glyphipterix arizonensis** Heppner, male (holotype). Madera Cyn., Santa Rita Mts., Santa Cruz Co., Arizona, 27–30 Jul 1947, L. M. Martin. (LA-CM) (slide JBH 1406) (see fig. 11, adult). 172. **Glyphipterix arizonensis** Heppner, male (holotype) aedeagus, same data.

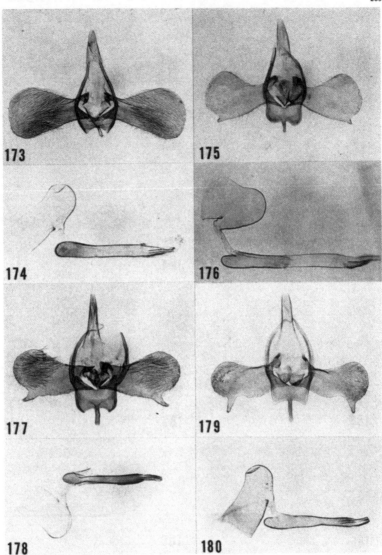

Figs. 173-180. Male genitalia. 173. **Glyphipterix roenastes** Heppner, male (holotype). Silverton, [San Juan Co.], Colorado, 16-23 Jul, [9200']. (USNM) (slide USNM 77138) (see fig. 90, adult). 174. **Glyphipterix roenastes** Heppner, male (holotype) aedeagus, same data. 175. **Glyphipterix californiae** Walsingham, male Mt. Shasta City, Siskiyou Co., California, 1 Jul 1970, P. Rude. (UCB) (slide JBH 529). 176. **Glyphipterix californiae** Walsingham, male aedeagus, same data. 177. **Glyphipterix chiri-cahuae** Heppner, male (holotype). S. W. R. S. [= Southwestern Res. Sta.], 5 mi. W. Portal, [Chiricahua Mts.], Cochise Co., Arizona, 9 Nov 1964, V. D. Roth. (AMNH) (slide JBH 443) (see fig. 102, adult). 178. **Glyphipterix chiricahuae** Heppner, male (holotype) aedeagus, same data. 179. **Glyphipterix hodgesi** Heppner, male (paratype). Fort Valley, 7.5 mi. NW. Flagstaff, [San Francisco Mts.], Coconino Co., Arizona, 23 Aug 1961, 7350', R. W. Hodges. (USNM) (slide USNM 7749). 180. **Glyphipterix hodgesi** Heppner, male (paratype) aedeagus, same data.

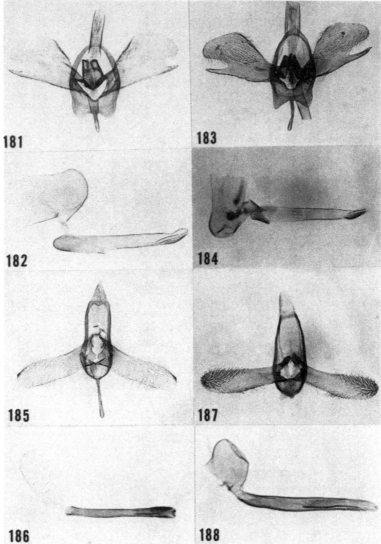

Figs. 181–188. Male genitalia. 181. **Glyphipterix montisella** Chambers, male (lectotype). [Spanish Bar, Fall River], Denver, South Park, Colorado, [V. T. Chambers]. (MCZ) (slide JBH 411. 182. **Glyphipterix montisella** Chambers, male aedeagus, same data. 183. **Glyphipterix santaritae** Heppner, male (paratype). Santa Catalina Mts., Pima Co., Arizona, 15 Jul 1938, Bryant. (LACM) (slide JBH 442). 184. **Glyphipterix santaritae** Heppner, male (paratype) aedeagus, same data. 185. **Glyphipterix haworthana** (Stephens), male. Mer Bleue, Ontario, 13 May 1932, J. H. McDunnough. (USNM) (slide USNM 77106). 186. Glyphipterix haworthana (Stephens), male aedeagus, same data. 187. **Glyphipterix sistes sistes** Heppner, male (paratype). Duncan, Vancouver Is., British Columbia, Canada, 31 Jul 1925, A. W. Hanham. (USNM) (slide USNM 77153). 188. **Glyphipterix sistes sistes** Heppner, male (paratype) aedeagus, same data.

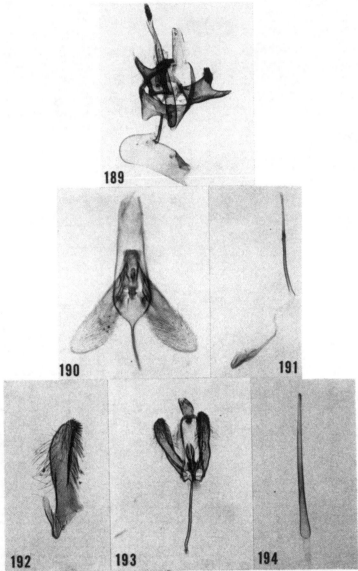

Figs. 189-194. Male genitalia. 189. **Glyphipterix ruidosensis** Heppner, male (holotype). Ruidoso Cyn. [Sacramento Mts., Otero-Lincoln Co.], New Mexico, 30 Sep 1916, C. Heinrich. (USNM) (slide USNM 77805) (see fig. 114, adult). Aedeagus in situ. 190. **Diploschizia regia** Heppner, male (holotype). Royal Palm State Pk. [= Royal Palm Hammock, Everglades Natl. Pk., Dade Co.], Florida, Jan 1930, F. M. Jones. (USNM) (slide USNM 77810) (see fig. 117, adult). 191. **Diploschizia regia** Heppner, male (holotype) aedeagus, same data. 192. **Diploschizia kimballi** Heppner, male (paratype) left valva (meso-lateral view showing ventral appendage). Archer Road Lab, 3 mi. SW. Gainesville, Alachua Co., Florida, 2 Apr 1976, J. B. Heppner. (JBH) (slide JBH 1461). 193. **Diploschizia kimballi** Heppner, male (holotype). Lake Panasoffkee, Sumter Co., Florida, 11 May 1974, emerged 14 May 1974, K. W. Knopf. (USNM) (slide USNM 77822) (see fig. 118, adult). 194. **Diploschizia kimballi** Heppner, male (holotype) aedeagus, same data.

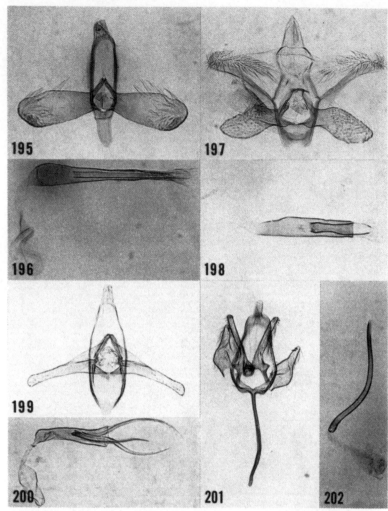

Figs. 195-202. Male genitalia. 195. **Diploschizia lanista** (Meyrick), male. Archbold Biological Sta., [10 mi. S.] Lake Placid, Highlands Co., Florida, 1-7 May 1964, R. W. Hodges. (USNM) (slide USNM 77145). 196. **Diploschizia lanista** (Meyrick), male aedeagus, same data. 197. **Diploschizia minimella** Heppner, male (holotype). Archbold Biological Sta., [10 mi. S.] Lake Placid, Highlands Co., Florida, 16-22 May 1964, R. W. Hodges. (USNM) (slide USNM 77156) (see fig. 122, adult). 198. **Diploschizia minimella** Heppner, male (holotype) aedeagus, same data. 199. **Diploschizia habecki** Heppner, male (holotype). Bivens Arm Lake, 3 mi. SW. Gainesville, Alachua Co., Florida, reared **ex Rhynchospora corniculata,** 5 Nov 1974, J. B. Heppner. (FSCA) (slide JBH 414). 200. **Diploschizia habecki** Heppner, male (holotype) aedeagus, same data. 201. **Diploschizia impigritella** (Clemens), male. Cincinnati, [Hamilton Co.], Ohio, 30 Jul 1907, A. F. Braun. (ANSP) (slide JBH 303). 202. **Diploschizia impigritella** (Clemens), male aedeagus, same data.

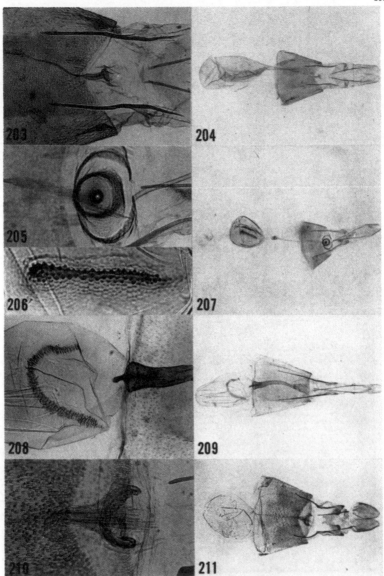

Figs. 203–211. Female Genitalia. 203. **Abrenthia cuprea** Busck, female ostium. Plummers Is., [Montgo-mery Co.], Maryland, 8 Jul 1968, P. J. Spangler. (USNM) (slide USNM 77763) (see fig. 153, adult). 204. **Abrenthia cuprea** Busck, female, same data. 205. **Neomachlotica spiraea** Heppner, female (paratype) ostium. Fisheating Cr., 2 mi. SE. Palmdale, Glades Co., Florida, 6 May 1975, on **Pluchea purpurascens** flowers, J. B. Heppner. (USNM) (slide USNM 77152) (see fig. 155, adult). 206. **Neo-machlotica spiraea** Heppner, female (paratype) signum, same data. 207. **Neomachlotica spiraea** Heppner, female (paratype), same data. 208. **Drymoana blanchardi** Heppner, female (paratype) signum. Camp Strake, Montgomery Co., Texas, 9 Sep 1975, A. & M. E. Blanchard. (USNM) (slide USNM 77150) (see fig. 157, adult). 209. **Drymoana blanchardi** Heppner, female (paratype), same data. 210. **Glyphipterix quadragintapunctata** Dyar, female ostium. Onaga, [Pottawatomie Co.], Kan-sas. (MCZ) (slide JBH 360). 211. **Glyphipterix quadragintapunctata** Dyar, female, same data.

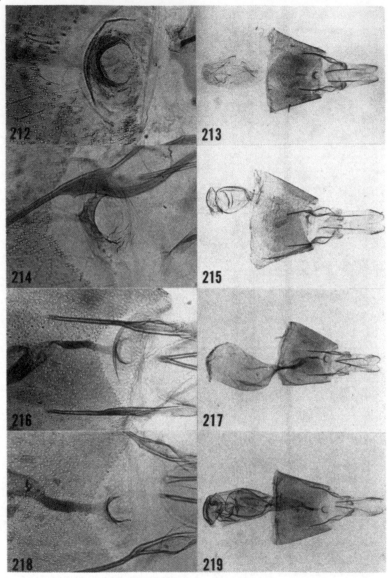

Figs. 212–219. Female Genitalia. 212. **Glyphipterix powelli powelli** Heppner, female (paratype) ostium. Inverness, Marin Co., California, 5 Apr 1959, D. Burdick. (UCB) (slide JBH 381) (see fig. 161, adult). 213. **Glyphipterix powelli powelli** Heppner, female (paratype), same data. 214. **Glyphipterix powelli jucunda** Heppner, female (paratype) ostium. Pullman, [Whitman Co.], Washington, 2 Jun 1965, R. D. Akre. (WSU) (slide JBH 523) (see fig. 163, adult). 215. **Glyphipterix powelli jucunda** Heppner, female (paratype), same data. 216. **Glyphipterix urticae urticae** Heppner, female (paratype) ostium. S. Fork Bonito Cr., Cpgd., Sacramento Mts., Lincoln Co., New Mexico, 6 Jul 1977, on Utica sp., J. B. Heppner. (USNM) (slide USNM 77784). 217. **Glyphipterix urticae urticae** Heppner, female (paratype), same data. 218. **Glyphipterix urticae sylviborealis** Heppner, female (paratype) ostium. Aweme, Manitoba, Canada, 30 Jun 1922, N. Criddle. (CNC) (slide JBH 1353). 219. **Glyphipterix urticae sylviborealis** Heppner, female (paratype), same data.

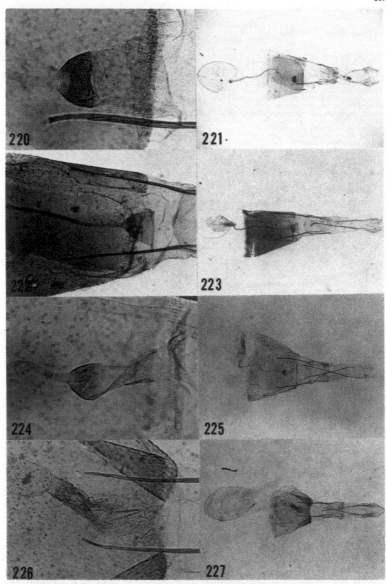

Figs. 220–227. Female Genitalia. 202. **Glyphipterix bifasciata** Walsingham, female ostium. 5 mi. E. McCloud, Siskiyou Co., California, 7 Jul 1957, J. A. Powell. (UCB) (slide JBH 377). 221. **Glyphipterix bifasciata** Walsingham, female, same data. 222. **Glyphipterix hypenantia** Heppner, female (paratype), ostium. 1 mi. NW Bartle, Siskiyou Co., California, 20 Jul 1966, P. Rude. (UCB) (slide JBH 1352). 223. **Glyphipterix hypenantia** Heppner, female (paratype), same data. 224. **Glyphipterix unifasciata** Walsingham, female ostium. Novato, Marin Co., California, 6 May 1962, D. C. Rentz. (CAS) (slide JBH 463) (see fig. 171, adult). 225. **Glyphipterix unifasciata** Walsingham, female, same data. 226. **Glyphipterix circumscriptella circumscriptella** Chambers, female ostium. Essex Co. Park, New Jersey, 1 Jul, W. D. Kearfott. (USNM) (slide USNM 77136) (see fig. 175, adult). 227. **Glyphipterix circumscripta circumscriptella** Chambers, female, same data.

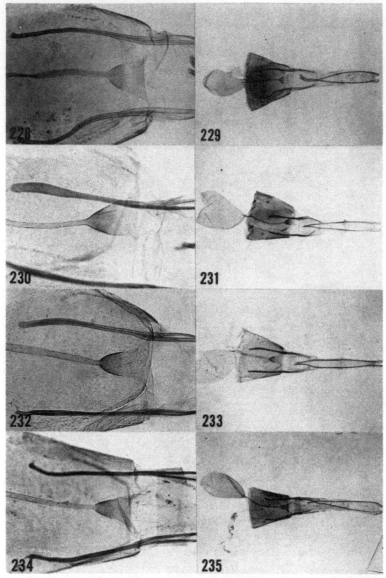

Figs. 228–235. Female Genitalia. 228. **Glyphipterix juncivora** Heppner, female (paratype) ostium. Lake Cpgd., 7 mi. E. Ephraim, Sanpete Co., Utah, 10 Jul 1976, 8500', on **Juncus**, J. B. Heppner. (USNM) (slide USNM 77791). 229. **Glyphipterix juncivora** Heppner, female (paratype), same data. 230. **Glyphipterix feniseca** Heppner, female (paratype) ostium. Alpine Lake, Marin Co., California, 25 Apr 1958. J. A. Powell. (UCB) (slide JBH 1402) (see fig. 97, adult). 231. **Glyphipterix feniseca** Heppner, female (paratype), same data. 232. **Glyphipterix sierranevadae** Heppner, female (paratype) ostium. Mineral King, Tulare Co., California, 24–31 Jul. (USNM) (slide USNM 77796) (see fig. 99, adult). 233. **Glyphipterix sierranevadae** Heppner, female (paratype), same data. 234. **Glyphipterix arizonensis** Heppner, female (paratype) ostium. "Arizona", 1882, Morrison, Walsingham Coll. 35562. (BMNH) (slide BM 20272) (see fig. 101, adult). 235. **Glyphipterix arizonensis** Heppner, female (paratype), same data.

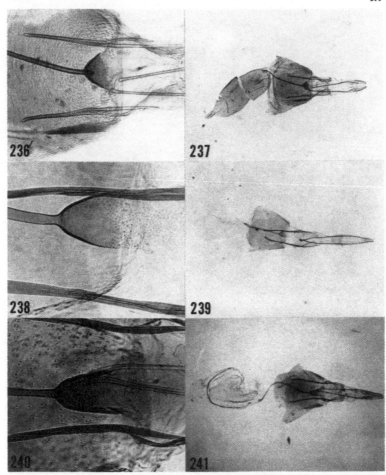

Fig. 236–241. Female Genitalia. 236. **Glyphipterix roenastes** Heppner, female (paratype) ostium. 15 mi. SE. Heber City, Wasatch Co., Utah, 7 Jul 1976, J. B. Heppner. (USNM) (slide USNM 77786) (see fig. 91, adult). 237. **Glyphipterix roenastes** Heppner, female (paratype), same data. 238. **Glyphipterix californiae** Walsingham, female ostium. Mt. Shasta City, Siskiyou Co., California, 1 Jul 1970, P. Rude. (UCB) (slide JBH 530) (see fig. 93, adult). 239. **Glyphipterix californiae** Walsingham, female, same data. 240. **Glyphipterix chiricahuae** Heppner, female (paratype) ostium. S. W. R. S. [= Southwestern Res. Sta.], 5 mi. W. Portal, [Chiricahua Mts.], Cochise Co., Arizona, 9 Nov 1964, V. D. Roth. (USNM) (slide USNM 77158). Accessory bursa pressed to bursa. 241. **Glyphipterix chiricahuae** Heppner, female (paratype), same data.

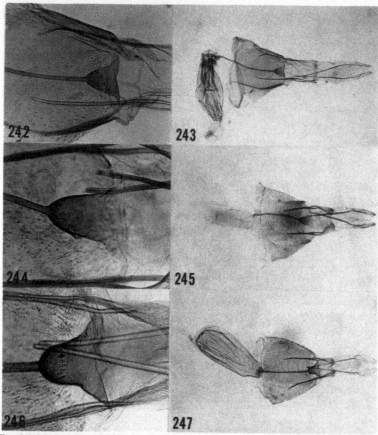

Figs. 242–247. Female Genitalia. 242. **Glyphipterix hodgesi** Heppner, female (paratype), ostium. Hart Prarie, 10 mi. NW. Flagstaff, [San Francisco Mts.], Coconino Co., Arizona, 28 Aug 1961, R. W. Hodges. (USNM) (slide USNM 77748) (see fig. 109, adult). 243. **Glyphipterix hodgesi** Heppner, female (paratype), same data. 244. **Glyphipterix montisella** Chambers, female ostium. Hidden Valley, Rocky Mtn. Natl. Pk., Colorado, 11 Aug 1929, A. F. Braun. (ANSP) (slide JBH 407). 245. **Glyphipterix montisella** Chambers, female, same data. 246. **Glyphipterix santaritae** Heppner, female (paratype) ostium. Bear Cyn., Santa Catalina Mts., [Pima Co.], Arizona, 2 Aug 1970, J. A. Powell. (UCB) (slide JBH 448) (see fig. 113, adult). 247. **Glyphipterix santaritae** Heppner, female (paratype), same data.

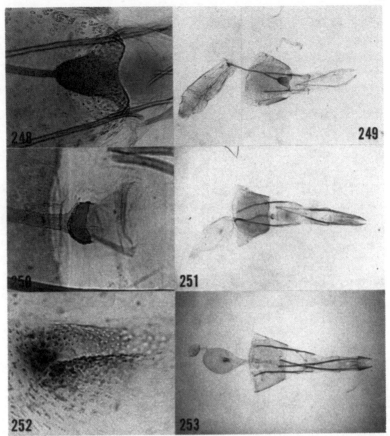

Figs. 248–253. Female Genitalia. 248. **Glyphipterix saurodonta** Meyrick, female ostium. Lost River St. Pk., [Hardy Co.], West Virginia, 19 Sep 1938, A. F. Braun. (ANSP) (slide JBH 439). 249. **Glyphipterix saurodonta** Meyrick, female, same data. 250. **Glyphipterix haworthana** (Stephens), female ostium. Mer Bleue, Ontario, Canada, 13 May 1932, G. S. Walley. (USNM) (slide USNM 77202). 251. **Glyphipterix haworthana** (Stephens), female, same data. 252. **Glyphipterix sistes sistes** Heppner, female (holotype) signum. Duncan, Vancouver Is., British Columbia, Canada, 31 Jul 1925, A. W. Hanham. (USNM) (slide USNM 77154) (see fig. 87, adult). 253. **Glyphipterix sistes sistes** Heppner, female (holotype), same data.

214

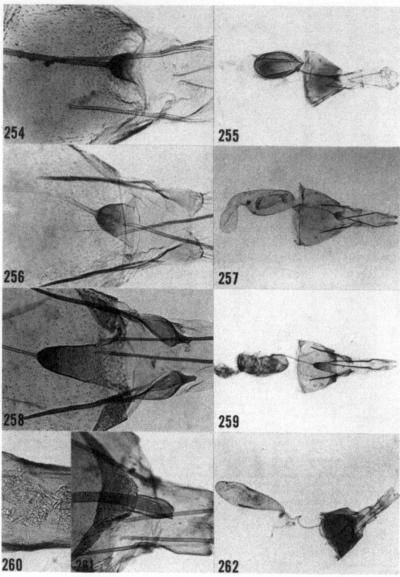

Figs. 254-262. Female Genitalia. 254. **Glyphipterix cherokee** Heppner, female (holotype) ostium. Great Smoky Mts., Tennessee, 23 Aug 1950, 6000', G. S. Walley. (CNC) (slide JBH 1410) (see Fig. 105, adult). 255. **Glyphipterix cherokee** Heppner, female (holotype), same data. 255. **Glyphipterix chambersi** Heppner, female (paratpye) ostium. "Kentucky", [no date], Chambers. (USNM) (slide USNM 77799). 257. **Glyphipterix chambersi** Heppner, female (paratype), same data. 258. **Glyphipterix ruidosensis** Heppner, female (paratype) ostium. Ruidoso Cyn., [Sacramento Mts., Otero-Lincoln Co.], New Mexico, 4 Oct 1916, C. Heinrich. (USNM) (slide USNM 77804) (see fig. 115, adult). 259. **Glyphipterix ruidosensis** Heppner, female (paratype), same data. 260. **Diploschizia kimballi** Heppner, female (paratype) signum. Archer Road Lab, 3 mi. SW. Gainesville, Alachua Co., Florida, 2 Apr 1976, J. B. Heppner. (JBH) (slide JBH 1472). 261. **Diploschizia kimballi** Heppner, female (paratype) ostium, same data. 262. **Diploschizia kimballi** Heppner, female (paratype), same data.

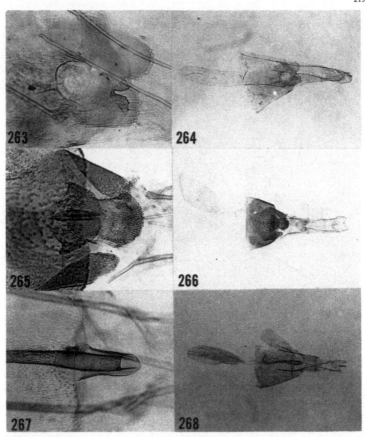

Figs. 263–268. Female Genitalia. 263. **Diploschizia minimella** Heppner, female (paratype) ostium. Lake Placid, Highlands Co., Florida, 30 Apr 1964, R. W.Hodges. (USNM) (slide USNM 77157). 264. **Diploschizia minimella** Heppner, female (paratype), same data. 265. **Diploschizia habecki** Heppner, female (paratype) ostium. Bivens Arm Lake, 3 mi. SW. Gainesville, Alachua Co., Florida, reared **ex Rhynchospora corniculata**, 17 Aug 1974, D. H. Habeck. (FSCA) (slide JBH 415). 266. **Diploschizia habecki** Heppner, female (paratype), same data. 267. **Diploschizia impigritella** (Clemens), female ostium. Cincinnati, [Hamilton Co.], Ohio, 9 Jun 1907, A. F. Braun. (USNM) (slide USNM 77302). 268. **Diploschizia impigritella** (Clemens), female same data.

216

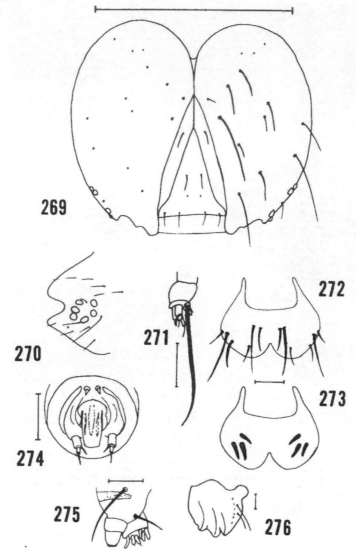

Figs. 269–276. Larva of Glyphipterix semiflavana Issiki. (after Moriuti, 1960). 269. Head chaetotaxy (scale line = 1 mm). 270. Lateral view of stemmata. 271. Ventral view of right antenna (scale line = 0.05 mm). 272. Labrum, dorsal view (scale line = 0.05 mm). 273. Labrum, ventral view. 274. Prementum apex, ventral view (scale line = 0.05 mm). 275. Left maxilla, ventral view (scale line = 0.05 mm). 276. Right mandible, ventral view (scale line = 0.05 mm).

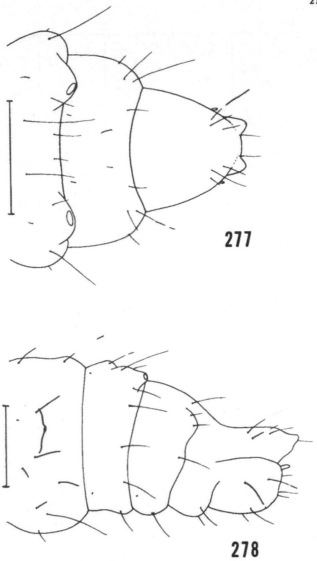

Figs. 277-278. Larva of **Glyphipterix semiflavana** Issiki. (after Moriuti, 1960). 277. Dorsal view of abdominal segments 8-10 (scale line = 0.5 mm). 278. Lateral view of abdominal segments 7-10.

218

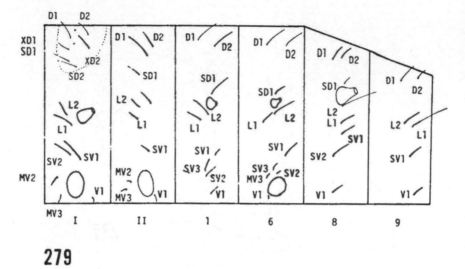

279

Fig. 279. Chaetotaxy of **Glyphipterix semiflavana** Issiki (after Moriuti, 1960).

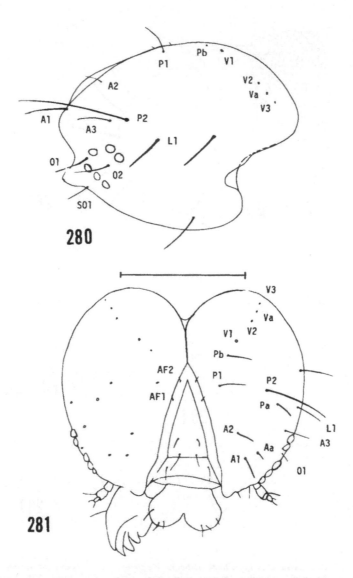

Figs. 280–281. Larva of **Diploschizia habecki** Heppner. 280. Head, lateral view. 281. Head, frontal view (scale line = 0.2 mm).

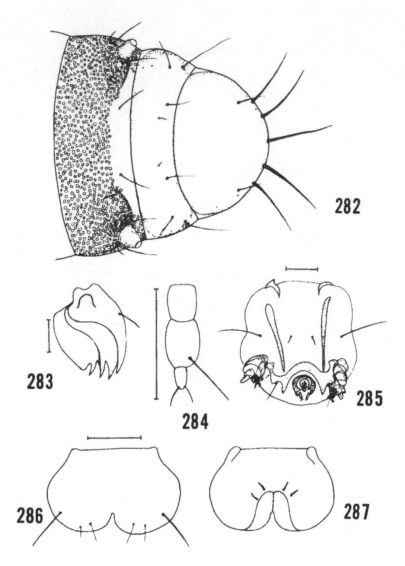

Figs. 282-287. Larva of **Diploschizia habecki** Heppner. 282. Abdominal tergites 8-10. 283. Right mandible, ventral view (scale line = 0.05 mm). 284. Left antenna, dorsal view (scale line = 0.05 mm). 285. Submentum, ventral view (scale line = 0.05 mm). 286. Labrum, dorsal view (scale line = 0.05 mm). 287. Labrum, ventral view.

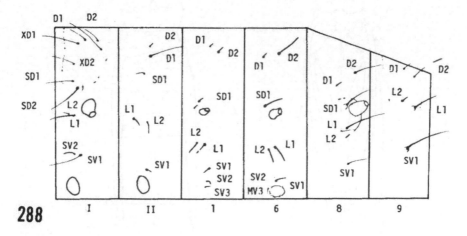

288

Fig. 288. Chaetotaxy of **Diploschizia habecki** Heppner.

222

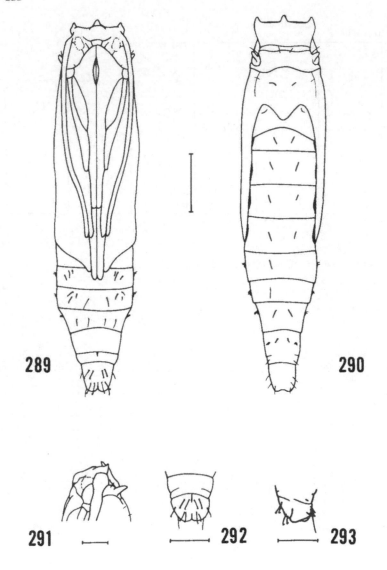

289   290

291 ⊢——⊣   ⊢——⊣ 292   ⊢——⊣ 293

Figs. 289–293. Pupa of **Glyphipterix semiflavana** Issiki. (after Moriuti, 1960).
289. Pupa, ventral view (scale line = 1 mm). 290. Pupa, dorsal view. 291.
Head, lateral view (scale line = 0.5 mm). 292. Posterior sternites, showing
hook-tipped setae (sale line = 0.5 mm). 293. Posterior end of abdomen,
lateral view (scale line = 0.5 mm).

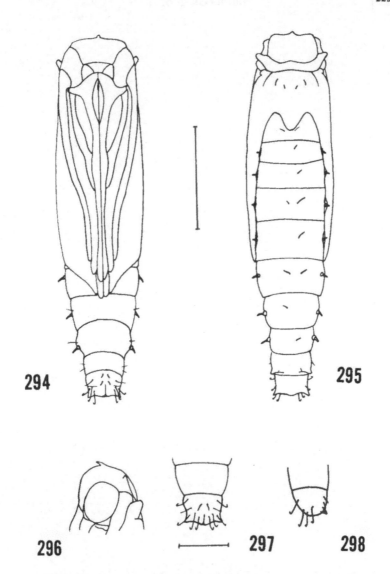

Figs. 294–298. Pupa of Diploschizia habecki Heppner. 294. Pupa, ventral view (scale line = 1 mm). 295. Pupa, dorsal view. 296. Head, lateral view (scale line = 0.5 mm). 297. Posterior sternites, showing hook-tipped setae. 298. Posterior end of abdomen, lateral view.

REFERENCES

Alford, D. V.
  1971. Synonymy as a source of confusion in distinguishing between certain tineid Lepidoptera. Plant Pathol., 20:171-173, pl. 4.
Amsel, H. G.
  1930. Die Mikrolepidopterenfauna der Mark Brandenburg nach dem heutigen Stande unserer Kenntnisse. Dtsch. Ent. Zeit. Iris, 44:83-132.
  1932. Neue mitteleuropäische Kleinschmetterlinge und Bemerkungen über **Melasina lugubris** Hb., und **M. ciliaris** O. (Lep.). Dtsch. Ent. Zeit. Iris, 46:18-24, 1 pl.
  1936. Zur Kenntnis der Kleinschmetterlingsfauna Sardiniens. Veröff. Dtsch. Übersee Mus. (Bremen), 1:344-365, 1 pl.
  1938. Neue oder seltene Kleinschmetterlinge aus dem Nordwestdeutschen Faunengebiet, nebst Bemerkungen über die Rassenbildung bei **Gelechia ericetella** Hb. in Mittel- und Südeuropa (Lepidoptera: Gelechiidae, Glyphipterygidae). Abh. Naturwiss. Ver. Bremen, 30:108-114.
  1949. Eine neue deutsche **Glyphipteryx**-Art. (Lep. Glyphipterygidae). Entomon, 1:88-89.
  1950. Kleinschmetterlinge aus dem badischen Schwarzwald. Beitr. Naturk. Forsch. SW-Dtsch., 9:26-28.
  1959. Portugiesische Kleinschmetterlinge gesammelt von Teodoro Monteiro, O.S.B. Ann. Facul. Cienc. Porto, 41:1-20, 2 pls.
Anderson, E. M.
  1904. Catalogue of British Columbia Lepidoptera. Victoria: Provincial Mus. 56 pp.
Anon.
  1897. Glyphipterygidae. In, The century dictionary and cyclopedia. Vol. 3:2551. New York: Century Co.
Bang-Haas, A.
  1875. Fortegnelse over de i Danmark levende Lepidoptera. Naturhist. Tids., 3(9):377-567; 3(10):1-56.
  1881. Fortegnelse over de i Danmark levende Lepidoptera. (Suppl.). Naturhist. Tids., 3(13):167-227.
Bankes, E. R.
  1907. **Glyphipteryx thrasonella**, Scop., ab. nitens n. ab. Ent. Mon. Mag., 43:204.

Barnes, W., and J. H. McDunnough
  1917. Check list of the Lepidoptera of boreal America.
        Decatur. 392 pp.
Barrett, C. G.
  1874. Fauna and flora of Norfolk.   Part V. Lepidoptera.
        Trans. Norfolk Norwich Nat. Soc., 1(suppl.):1-80.
Bauer, E.
  1923. Eine neue Glyphipteryx (Microlep.) aus der Alpen.
        Ent. Mitt., 12:167-168.
Bleszynski, S., J. Razowski, and R. Zukowski
  1965. [The lepidopterous fauna of the Pieniny Mountains].
        Acta Zool. Cracov., 10:375:493. [In Polish].
Boisduval, J. B. A. D. de
  1836. Histoire naturelle des insectes. Species général des
        Lépidoptères. Vol. 1. Paris: Roret. 690 pp.
Börner, C.
  1939. Die Grundlagen meines Lepidopteren-Systems. Proc.
        7th Internatl. Congr. Ent. 1938, 2:1372-1424.
  1953. Lepidoptera. In, P. Brohmer, Fauna von Deutsch-
        land, 382-421. Leipzig.
Bourgogne, J.
  1951. Ordre des Lépidoptères. In, P. P. Grassé, Traité de
        Zoologie. Anatomie, systématique, biologie. Tome
        X. Insectes supérieurs et Hémiptéroîdes. (Premier
        fascicule), 174-448. Paris: Masson & Editeurs. 975
        pp., 5 pls.
Bradley, J. D.
  1965. Microlepidoptera. In, Ruwenzori expedition 1952, 2
        (12):81-148. London: British Mus. (Nat. Hist.).
  1972. [Microlepidoptera]. In, J. D. Bradley, D. S. Fletcher,
        and P. E. S. Whalley, Part 2: Lepidoptera, in, G. S.,
        Kloet and W. D. Hincks, A check list of British in-
        sects, sec. ed. (rev.), 1-48. London: Roy. Ent. Soc.
        Lond. 153 pp.
Bradley, J. D., and E. C. Pelham-Clinton
  1967. The Lepidoptera of the Burren, Co. Clare, W. Ire-
        land. Ent. Gaz., 18:115-153, pl. 5.
Bradley, J. D., and K. Sattler
  1978. Comments on the proposed use of the plenary pow-
        ers to designate a type-species for the nominal ge-
        nus Glyphipterix Hübner, [1825] (Lepidoptera, Gly-
        phipterigidae). Z. N. (S.) 2115. Bull. Zool. Nomencl,
        35:71-73.
Braun, A. F.
  1924. The frenulum and its retinaculum in the Lepidoptera.
        Ann. Ent. Soc. Amer., 17:234-256, pl. 23.

1925. Microlepidoptera of northern Utah. Trans. Amer. Ent. Soc., 51:183-226.
1940. Notes and new species in the yponomeutoid group (Microlepidoptera). Trans. Amer. Ent. Soc., 66:273-282.

Brimley, C. S.
1938. The insects of North Carolina. Raleigh: North Carolina Dept. Agric. 560 pp.

Brock, J. P.
[1968]. The systematic position of the Choreutinae (Lep., Glyphipterygidae). Ent. Mon. Mag., 103:245-246.
1971. A contribution towards an understanding of the morphology and phylogeny of the ditrysian Lepidoptera. J. Nat. Hist., 5:29-102.

Bruand d'Uzelle, C. T.
1850. Catalogue systématique et synonymique des Lépidoptères du departement du Doubs. Tineides. Mém. Soc. Emul. Doubs, 3(3):23-68.

Busck, A.
1903. Notes on Brackenridge Clemens' types of Tineina. Proc. Ent. Soc. Wash., 5:181-200.
1904. Tineid moths from British Columbia, with descriptions of new species. Proc. U. S. Natl. Mus., 27:745-778.
1914. New genera and species of Microlepidoptera from Panama. Proc. U. S. Natl. Mus., 47:1-67.
1915. Descriptions of new North American Microlepidoptera. Proc. Ent. Soc. Wash., 17:79-94.
[1934]. Microlepidoptera of Cuba. Ent. Amer., 13:151-202, pl. 30-36.

Caradja, A. von
1899. Zusammenstellung der bisher in Rumänien beobachteten Microlepidopteren. Dtsch. Ent. Zeit. Iris, 12: 171-218.

Chambers, V. T.
1875a. Teneina [sic] of the United States. Cincinnati Qtr. J. Sci., 2:226-259.
1875b. Teneina [sic] of Colorado. Cincinnati Qtr. J. Sci., 2:289-305.
1876. Tineina. Can. Ent., 8:217-220.
1877a. Tineina. Can. Ent., 9:13-15.
1877b. The Tineina of Colorado. Bull. U. S. Geol. Geog. Survey Terr., 3:121-142.
1877c. Notes on a collection of tineid moths made in Colorado in 1875 by A. S. Packard, Jr., M.D. Bull. U. S. Geol. Geog. Survey Terr., 3:142-145.

1877d. On the distribution of Tineina in Colorado. Bull. U. S.
    Geol. Geog. Survey Terr., 3:147-150.
1878a. Tineina and their food-plants. Bull. U. S. Geol. Geog.
    Survey Terr., 4:107-123.
1878b. Index to the described Tineina of the United States
    and Canada. Bull. U. S. Geol. Geog. Survey Terr.,   4:
    124-167.
1878c. On Pronuba yuccasella (Riley), and the habits of some
    Tineina. J. Cincinnati Soc. Nat. Hist., 1:141-154.
1880a. Notes upon some tineid larvae. Psyche, 3:63-68.
1880b. Illustrations of the neuration of the wings of American
    Tineina. J. Cincinnati Soc. Nat. Hist., 2:194-204.
1881. New species of Tineina. J. Cincinnati Soc. Nat. Hist.,
    4:289-296.
Chinery, M.
1972. A field guide to the insects of Britain and northern Eu-
    rope. London: Collins. 352 pp., 60 pls.
Christoph, H.
1882. Neue Lepidopteren des Amurgebietes.   Bull. Soc. Imp.
    Natur. Moscou, 57:5-47.
Chopra, R. L.
1925. On the structure, life-history, economic importance and
    distribution of the cooksfoot moth, Glyphipteryx fisch-
    eriella, Zell. Ann. Appl. Biol., 12:359-397.
Clarke, J. F. G.
1941. The preparation of slides of the genitalia of Lepidop-
    tera. Bull. Brooklyn Ent. Soc., 36:149-161.
1955. Catalogue of the type specimens of microlepidoptera in
    the British Museum (Natural History) described by Ed-
    ward Meyrick. Vol. 1. London: British Mus. (Nat. Hist.).
    332 pp.
1969. Catalogue of the type specimens of microlepidoptera in
    the British Museum (Natural History) described by Ed-
    ward Meyrick. Vol. VI. Glyphipterygidae, Gelechiidae
    (A-C). London: British Mus. (Nat. Hist.). 537 pp.
Clemens, B.
1863. American micro-lepidoptera.   Proc. Ent. Soc. Phila.,
    2:4-14.
Common, I. F. B.
1966. Australian moths. (rev. ed.). Brisbane: Jacaranda Press.
    129 pp.
1970a. Lepidoptera: Yponomeutidae of Heard Island. Pacific
    Ins. Monog., 23:229-233.
1970b. Lepidoptera (moths and butterflies). In, The insects
    of Australia. A textbook for students and research
    workers. Carlton: Melbourne Univ. Press. 1029 pp. (Pp
    765-866).

1974. Lepidoptera (moths and butterflies). In, The insects of Australia. A textbook for students and research workers. Supplement 1974. Carlton: Melbourne Univ. Press. 146 pp. (Pp. 98-107).

1975. Evolution and classification of the Lepidoptera. Ann. Rev. Ent., 20:183-203.

Comstock, J. H.
1924. An introduction to entomology. Ithaca: Comstock. 1044 pp.

1940. An introduction to entomology. (9th rev. ed.). Ithaca: Comstock. 1064 pp.

Costa Lima, A. da
1945. Insetos do Brasil. Lepidopteros. Vol. 5, pt. 1. Escola Noc. Agron., Ser. Didatica, 7:1-379.

Covell, C. V., Jr.
1984. A field guide to the moths of eastern North America. Boston: Houghton Mifflin. 496 pp., 64 pl.

Crombrugghe de Picquerdaele, G. de
1906. Catalogue raisonné des microlépidoptères de Belgique. Deuxième partie. Mém. Soc. Ent. Belg., 14:1-155.

Curo, A.
1883. Saggio di un catalogo dei Lepidotteri d'Italia. Bull. Soc. Ent. Ital., 15:1-144.

Curtis, J.
[1824]-40. British entomology; being illustrations and descriptions of the genera of insects found in Great Britain and Ireland: containing coloured figures from nature of the most rare and beautiful species, and in many instances of the plants upon which they are found. London. 16 vols., 770 pls. [1827: Vol. 6. Lepidoptera, Pt. 2].

Davies, T. H.
1973. List of Lepidoptera collected in areas surrounding Hastings and Napier. New Zealand Ent., 5:204-216.

Davis, D. R.
1967. A revision of the moths of the subfamily Prodoxinae (Lepidoptera: Incurvariidae). Bull. U. S. Natl. Mus., 255:1-170.

Desmarest, E.
[1857]. Papillons nocturnes. In, J. C. Chenu, Encyclopédie d'histoire naturelle ou traité complet de cette science d'apres les travaux des naturalists les plus éminents de tous les pays et de toutes époques. Dixième Tome. Paris: Havard. 312 pp., 39 pl.

Diakonoff, A. N.
1948. Fauna Buruana. Microlepidoptera II. Treubia, 19:197-219.

1950. A revision of the family Ceracidae (Lepidoptera Tortricoidea). Bull. Br. Mus. (Nat. Hist.), Ent., 1:171-219.

[1968]. Microlepidoptera of the Philippine Islands. Bull. U. S. Natl. Mus., 257:1-484.

1976. Aantekeningen over de Nederlandse Microlepidoptera 3 (Glyphipterigidae). Ent. Ber., 36:82-84.

1977a. Description of Hilarographini, a new tribus in the Tortricidae (Lepidoptera). Ent. Ber., 37:76-77.

1977b. Synonymic notes on the Glyphipterygidae, with selection of neotypes for two Phyllonorycter species, Lithocolletidae (Lepidoptera). Zool. Meded., 51:171-176.

1977c. Rearrangement of certain Glyphipterygidae sensu Meyrick, 1913, with descriptions of new taxa (Lepidoptera). Zool. Verh., 158:1-55.

1978. Descriptions of new genera and species of the so-called Glyphipterygidae sensu Meyrick, 1913 (Lepidoptera). Zool. Verh., 160:1-63.

1979. Descriptions of new taxa of the so-called Glyphipterigidae auctorum (Lepidoptera). Zool. Meded., 54:291-312.

(in press). Glyphipterigidae sensu lato. In, H. G. Amsel, et al. (eds.), Microlepidoptera Palaearctica. Vol. 7. Karlsruhe: G. Braun.

Diakonoff, A. N., and Y. Arita
1976. New and old species of Glyphipterigidae and a new species of Acrolepiidae from Japan and the Far East (Lepidoptera). Zool. Meded., 50:179-219.

Diakonoff, A. N., and J. B. Heppner
1977. Proposed use of the plenary powers to designate a type-species for the nominal genus Glyphipterix Hübner, [1825] (Lepidoptera, Glyphipterygidae). Z. N. (S.) 2115. Bull. Zool. Nomencl., 34:81-84.

Donckier de Donceel, C.
1882. Catalogue des lépidoptères de Belgique. Ann. Soc. Ent. Belg., 26:5-161.

Duckworth, W. D.
1971. Neotropical microlepidoptera XX: revision of the genus Setiostoma (Lepidoptera: Stenomidae). Smithson. Contr. Zool., 106:1-45.

Duckworth, W. D., and T. D. Eichlin
1977. A classification of the Sesiidae of America north of Mexico (Lepidoptera: Sesioidea). Occas. Pap. Ent. (Sacramento), 26:1-54.

Dugdale, J. S.
1975. The insects in relation to plants. In, G. Kuschel (ed.), Biogeography and ecology in New Zealand. The Hague: W. Junk. 680 pp. (Pp. 561-589).

Duponchel, P. A. J.
  1838. Nocturnes. In, J. B. Godart (ed.), Histoire naturelle des
    lépidoptères ou papillons de France. Vol. 11.   Paris:
    Méquignon-Marvis. 720 pp., pl. 287-314.
  1841. Aechmia. In, C. d'Orbigny (ed.), Dictionnaire universel
    d' histoire naturelle résumnant et completant tour les
    faits présentés par les encyclopédies les anciens dict-
    ionnaires scientifiques les oeuvres completes de Buffon,
    et les traites spéciaux sur les diverses branches des sci-
    ences naturelles donnant la description des etres et des
    divers phénomènes de la nature. Vol. 1:36.   Paris: Re-
    nard. 649 pp.
  [1845]. Catalogue méthodique des lépidoptères d' Europe dis-
    tribués en familles, tribus et genres, avec l' exposé des
    caractères sur lesquels ces divisions sont fondées et l'in-
    dication des lieux et des époques où l' on trouve chaque
    espèce, pour servir de complément et de rectification
    a l' histoire naturelle des lépidoptères l' Europe par les
    suppléments qu'on y a ajoutés. Paris: Méquignon-Mar-
    vis. 523 pp.
  1845. Glyphipteryx. In, C. d' Orbigny (ed.), op cit. Vol. 6:243.
    Paris: Renard. 792 pp.
Dyar, H. G.
  1900. Notes on some North American Yponomeutidae.   Can.
    Ent., 32:37-41.
  [1903]. A list of North American Lepidoptera and key to the
    literature of this order of insects.   Bull. U. S. Natl.
    Mus., 52:1-723.
Eckstein, K.
  1933. Die Schmetterlinge Deutschlands mit besonderer Be-
    rücksichtigung ihrer Biologie und wirtschaftlichen Be-
    deutung. 5 Band. Die Kleinschmetterlinge Deutschl-
    ands. Stuttgart: Lutz. 223 pp., 32 pl.
Empson, H.
  1956. Cocksfoot moth investigations.   Plant Pathol., 5:12-
    18, pl. 1.
Engel, H.
  1908. A preliminary list of the Lepidoptera of western Penn-
    sylvania collected in the vicinity of Pittsburgh.   Ann.
    Carnegie Mus., 5:27-136.
Esaki, T., et al.
  1932. Nippon Konchu Zukan. Iconographia insectorum Japon-
    icorum. Tokyo: Hokuryukan. 2 vol., 24 pl.
Essig, E. O.
  1941. Itinerary of Lord Walsingham in California and Oregon,
    1871-1872. Pan-Pac. Ent., 17:97-112, 1 pl.

Felder, R., and A. F. Rogenhofer
  1875. Zeite Abtheilung: Lepidoptera. Atlas. In, C. Felder, R.
    Felder, and A. F. Rogenhofer, Reise der österreichisch-
    en Fregatte Novara um die Erde in den Jahren 1857,
    1858, 1859 unter den Behilfen des Commodore B. von
    Wüllersstorf-Urbair. Zoologischer Theil. Zweiter Band.
    Vienna. 20 pp., pl. 108-140.
Fereday, R. W.
  [1898]. A synonymic list of the Lepidoptera of New Zealand.
    Trans. New Zealand Inst., 30:326-377.
Ferro, D. N., A. D. Lowe, R. G. Ordish, K. G. Somerfield, and J.
    C. Watt
  1977. Standard names for common insects of New Zealand.
    Bull. Ent. Soc. New Zealand, 4:1-42.
Fettig, A.
  1882. Microlépidoptères. In, H. Peyerimhoff, Catalogue des
    lépidoptères d'Alsace. Deuxième partie.   Bull. Soc.
    Hist. Nat. Colmar, 22/23:34-214.
Fibiger, M., and N. P. Kristensen
  1974. The Sesiidae (Lepidoptera) of Fennoscandia and Den-
    mark. Fauna Ent. Scandinavica, 2:1-91.
Fischer von Röslerstamm, J. E.
  1834-43. Abbildungen zur Berichtigung und Ergänzung der
    Schmetterlingkunde, besonders der Microlepidopterolo-
    gie als Supplement zu Treitschke's und Hübner's euro-
    päischen Schmetterlingen, mit erläuterndem Text. Leip-
    zig. 308 pp., 100 pl. (20 pts.) [1841: 233-252, pl. 81-85
    (Heft 17)].
Fletcher, T. B.
  1927. Some Swiss micro-lepidoptera.   Ent. Rec. J. Var., 39:
    33-37.
  1929. A list of the generic names used for microlepidoptera.
    Mem. Dept. Agric. India, Ent. Ser., 11:1-246.
  1938. The study of microlepidoptera.   Ent. Rec. J. Var., 50:
    107-110.
  1946. Names of microlepidoptera. Ent. Rec. J. Var., 58:127-
    129.
Fletcher, T. B., and C. G. Clutterbuck
  [1941]. Microlepidoptera of Gloucestershire. [Pt. 4]. Proc.
    Cotteswold Naturalists' Field Club, 27:93-108.
Forbes, W. T. M.
  1923. The Lepidoptera of New York and neighboring states.
    Primitive forms: microlepidoptera, pyraloids, bomby-
    ces. Cornell Univ. Agric. Exp. Sta. Mem., 68:1-729.
Ford, L. T.

1949. A guide to the smaller British Lepidoptera.   London:
        So. London Ent. Nat. Hist. Soc. 230 pp.
1954. The Glyphipterygidae and allied families. Proc. Trans.
        So. London Nat. Hist. Soc., 1952-53:90-99, pl. 8.
Fré, C. de
1858. Catalogue des microlépidoptères de la Belgique. Ann.
        Soc. Ent. Belg., 2:45-162.
Frey, H.
1880. Die Lepidopteren der Schweiz.   Leipzig: Engelmann.
        454 pp.
Frost, S. W.
1964. Insects taken in light traps at the Archbold Biological
        Station, Highlands County, Florida. Fla. Ent., 47:129-
        161.
1975. Third supplement to insects taken in light traps at the
        Archbold Biological Station, Highlands County, Florida.
        Fla. Ent., 58:35-42.
Glaser, L.
1887. Catalogus etymologicus Coleopterorum et Lepidopter-
        orum. Berlin: Friedländer. 396 pp.
Glick, P. A.
1965. Review of collections of Lepidoptera by airplane. J.
        Lepid. Soc., 19:129-137.
Goater, B.
1974. The butterflies and moths of Hampshire and the Isle of
        Wight (being an account of the whole of the Lepidop-
        tera). Faringdon: Classey. 439 pp.
Gozmany, L.
1954. Studies on microlepidoptera.   Ann. Hist.-Nat. Mus.
        Natl. Hung., 5:273-285.
Hampson, G. F.
1918. Some small families of the Lepidoptera which are not
        included in the key to the families in the catalogue of
        Lepidoptera Phalaenae, a list of families and subfami-
        lies of the Lepidoptera with their types, and a key to
        the families. Novit. Zool., 25:366-394.
Handlirsch, A.
1925. Lepidoptera. In, C. Schröder, Handbuch der Entomolo-
        gie. Band III. Geschichte, Literatur, Technik, Paläon-
        tologie, Phylogenie, Systematik. Jena: Fischer.   1201
        pp. (Pp. 852-941).
Hannemann, H. J., and E. Urbahn
1974. Lepidoptera-Schuppenflügler, Schmetterlinge. In, E.
        Stresemann, Exkursionsfauna für die Gebiete der DDR
        und der BRD. Insekten-Zeiter Halbband.   Wirbellose

II/2, 142-322. Berlin: Volk & Wissen Volkseigener. 476 pp.

Hartig, F.
1956. Prodromus dei Microlepidotteri della Venezia Tridentina e delle regioni adiacenti.   Stud. Trentini Sci. Nat., 33:89-148.

Hartig, F., and H. G. Amsel
1951. Lepidoptera Sardinica. Frag. Ent., 1:1-159.

Hartmann, A.
1880. Die Kleinschmetterlinge des europäischen Faunengebietes. Erscheinungszeit der Raupen und Falter, Nahrung und biologische Notizen. Munich: T. Ackermann. 182 pp.

Heath, J.
1976. The moths and butterflies of Great Britain & Ireland. Volume 1. Micropterigidae-Heliozelidae.   London: Blackwell Sci. Publ. 343 pp., 13 pl.

Heinemann, H. von
1870. Die Schmetterlinge Deutschlands und der Schweiz. Zweite Abtheilung. Kleinschmetterlinge. Band II. Die Motten und Federmotten. Braunschweig: Schwetschke. 825 pp. (Index: 102 pp.).

Hennig, W.
1950. Grundzüge einer Theorie der phylogenetischen Systematik. Berlin. 370 pp. (English trans., 1965: Phylogenetic systematics. Urbana: Univ. Illinois Press. 263 pp.)

Heppner, J. B.
1974. **Tortyra slossonia** collected at UV light on Key Largo, Florida (Glyphipterygidae). J. Lepid. Soc., 28:292.

1977. The status of the Glyphipterigidae and a reassessment of relationships in yponomeutoid families and ditrysian superfamilies. J. Lepid. Soc., 31:124-134.

1978. Transfers of some Nearctic genera and species of Glyphipterigidae (auctorum) to Oecophoridae, Copromorphidae, Plutellidae, and Tortricidae (Lepidoptera). Pan-Pac. Ent., 54:48-55.

1979. Brachodidae, a new family name for Atychiidae (Lepidoptera: Sesioidea). Ent. Ber., 39:127-128.

1981a. Revision of the new genus **Diploschizia** (Lepidoptera: Glyphipterigidae) for North America.   Fla. Ent., 64: 309-336.

1981b. **Neomachlotica**, a new genus of Glyphipterigidae (Lepidoptera). Proc. Ent. Soc. Wash., 83:479-488.

1982a. Synopsis of the Glyphipterigidae (Lepidoptera: Copromorphoidea) of the world. Proc. Ent. Soc. Wash., 84: 38-66.

1982b. A world catalog of genera associated with the Glyph-
    ipterigidae auctorum (Lepidoptera). J. New York Ent.
    Soc., 89:220-294.
1982c. Synopsis of the Hilarographini (Lepidoptera: Tortrici-
    dae) of the world. Proc. Ent. Soc. Wash., 84:704-715.
1982d. Review of the family Immidae, with a world checklist
    (Lepidoptera: Immoidea). Entomography (Sacramento),
    1:257-279.
1983a. Glyphipterigidae. In, R. W. Hodges, et al.(eds.), Check
    list of the Lepidoptera of America north of Mexico.
    London: E. W. Classey. 284 pp. [Pp. 25-26].
1983b. Ecological notes on Brachodidae of eastern Europe.
    Nota Lepid. (Karlsruhe), 6:99-110.
1984. Glyphipterigidae. In, J. B. Heppner (ed.), Atlas of Ne-
    otropical Lepidoptera. Checklist: Part 1. Micropteri-
    goidea - Immoidea. The Hague: W. Junk. 112 pp. [Pp.
    54-55].
(in prep.). Sesioidea. Choreutidae. In, R. B. Dominick, et al.,
    The moths of America north of Mexico.   Fascicle 8.2.
    London: E. W. Classey.
Heppner, J. B., and W. D. Duckworth
1981. Classification of the superfamily Sesioidea (Lepidop-
    tera: Ditrysia). Smithson. Contr. Zool., 314:1-144.
Herbulot, C.
1949. Atlas des lépidoptères de France, Belgique, Suisse. III.
    Héterocères (fin). Paris: N. Boubee. 145 pp., 12 pl.
Hering, E. M.
1927. Die Minenfauna der canarischen Inseln.   Zool. Jahrb.,
    Abt. Syst. Ökolog. Geog. Tiere, 53:405-486.
1932. Die Schmetterlinge nach ihren Arten dargestellt.Bd. 1.
    In, P. Brohmer, et al., Die Tierwelt Mitteleuropas. Ein
    Handbuch zu ihrer Bestimmung als Grundlage für faun-
    istische-zoogeographische Arbeiten.   Leipzig: Von
    Quelle & Meyer. 545 pp.
1957. Bestimmungstabellen der Blattminen von Europa. The
    Hague: W. Junk. 1185 pp., 221 pl. (3 vol.).
Herrich-Schäffer, G. A. W.
1853-55. Systematische Bearbeitung der Schmetterlinge von
    Europa, zugleich als Text, Revision und Supplement zu
    Jakob Hübner's Sammlung europäischer Schmetterlinge.
    Fünfter Band. Die Schaben und Federmotten. Regens-
    burg: G. J. Manz. 394 pp. (index 52 pp.).   [1853:1-72.
    1854:73-224. 1855:225-394. Index, pp. 1-52.].
Heslop, I. R. P.
1945. Check-list of the British Lepidoptera with the english

                name of each of the 2299 species. London. 35 pp.
Hinton, H. E.
    1946. On the homology and nomenclature of the setae of lep-
           idopterous larvae, with some notes on the phylogeny of
           the Lepidoptera. Trans. Roy. Ent. Soc. Lond., 97:1-37.
Hodges, R. W.
    1974. Gelechioidea. Oecophoridae. In, R. B. Dominick, et al.
           (eds.), The moths of America north of Mexico.   Fasc.
           6.2. London: E. W. Classey. 142 pp., 8 pl.
Hrubý, K.
    1964. Prodromus Lepidopter Slovenska. Bratislava: Vyd. Slov.
           Akad. Vied. 962 pp.
Hübner, J.
    1816-[25]. Verzeichniss bekannter Schmettlinge [sic]. Augs-
           burg. 431 pp. [[1825]:305-431].
Hudson, G. V.
    1928. The butterflies and moths of New Zealand. Wellington:
           Ferguson & Osborn. 386 pp., 62 pl.
    1939. A supplement to the butterflies and moths of New Zea-
           land. Wellington: Ferguson & Osborn. Pp. 387-481; pl.
           63-72.
Hutton, F. W.
    1883. Index faunae Novae Zealandiae. London: Dulau. 372 pp.
ICZN
    1984. Report on Glyphipterix Hübner, [1825] (Insecta, Lepid-
           optera): Z.N.(S.) 2115. Bull. Zool. Nomencl., 41:250-253.
Imms, A. D.
    1948. A general textbook of entomology, including the anat-
           omy, physiology, development and classification of in-
           sects. New York: Dutton. 727 pp.
Inoue, H.
    1954. Check list of the Lepidoptera of Japan.   Part 1: Mic-
           ropterigidae - Phaloniidae. Tokyo: Rikusuisha. 112 pp.
           [In Japanese].
Inoue, H., M. Okano, T. Shirozu, S. Sugi, and H. Yamamoto
    1959. Iconographia insectorum Japonicorum. Colore naturali
           edita. Volumen 1 (Lepidoptera).   Tokyo: Hokuryukan.
           284 pp., 184 pl. [In Japanese].
Issiki, S. T.
    1930. New Japanese and Formosan microlepidoptera.   Ann.
           Mag. Nat. Hist., (10) 6:422-431.
Jäckh, E.
    1942. Die Microlepidopteren-Fauna des rechtsseitigen Mitt-
           el-rheintales nebst Beschreibung von Borkhausenia mag-
           natella spec. nov. (Lep. Gelechiidae). Zeit. Wiener Ent.

Ver., 27:137-141, 173-176, 187-200, 216-221, 272-274, pl. 11.

Janse, A. J. T.
1917. Check-list of the South African Lepidoptera Heterocera. Pretoria: Transvaal Mus. 219 pp.
1925. Key to the families, sub-families and tribes of the Lepidoptera, with special references to the South African species. So. Afr. J. Sci., 22:318-345.

Jones, F. G. W., and M. G. Jones
1964. Pests of field crops. New York: St. Martin's Press. 406 pp.

Jordan, K.
1886. Die Schmetterlingsfauna nordwest-Deutschlands, insbesondere die lepidopterologischen Verhältnisse der Umgebung von Göttingen. Zool. Jahrb., Zeit. Syst. Geog. Biol. Thiere, Supplementheft 1:1-164.

Jourdheuille, M. C.
1883. Catalogue des lépidoptères du departement de l'Aube. Mém. Soc. Acad. l'Aube, 47:1-229.

Karsholt, O., and E. S. Nielsen
1976. Systematisk fortegnelse over Danmarks sommerfugle. Klampenborg: Scand. Sci. Press. 128 pp.

Kearfott, W. D.
1903. Tineoidea. In, J. B. Smith, Check list of the Lepidoptera of boreal America. (2nd ed.). Philadelphia: Amer. Ent. Soc. 136 pp.

Kimball, C. P.
1965. The Lepidoptera of Florida. An annotated checklist. Arthropods of Florida and neighboring land areas, 1:1-363 (26 pl.). Gainesville: Fla. Dept. Agric., Div. of Pl. Indus.

King, J. J. F. X.
1901. Microlepidoptera. In, G. F. S. Elliot, M. Laurie, and J. B. Murdoch, Fauna, flora and geology of the Clyde area. [Vol. 1]. Glasgow. 398 pp.

Klimesch, J.
1961. Ordnung Lepidoptera. I. Teil: Pyralidina, Tortricina, Tineina, Eriocraniina und Micropterygidae. In, H. Franz, Die Nordost-Alpen im Spiegel ihrer Landtierwelt. Eine Gebietsmonographie umfassend: Fauna, Faunengeschichte, Lebensgemeinschaften und Beeinflussung der Tierwelt durch den Menschen. Innsbruck: Universitätsverlag Wagner. 798 pp. (Pp. 481-789).
1968. Die Lepidopterenfauna Mazedoniens. IV. Microlepidoptera. Skopje: Prirodonaucen Muz. Skopje. 203 pp.

Kloet, G. S., and W. D. Hincks
1945. A check list of British insects. Stockport, England. 483 pp.

Kodama, T.
1961. The larvae of Glyphipterygidae (Lepidoptera) in Japan (I.). Osaka Fac. Agric., Ent. Lab., 6:35–45. [In Japanese].

Krogerus, H., M. Ophiem, M. von Schantz, I. Svensson, and N. L. Wolff
1971. Catalogus lepidopterorum Fenniae et Scandinaviae. Microlepidoptera. Helsinki. 40 pp.

Kuznetsov, V. I., and A. A. Stekolnikov
1976. Phylogenetic relationships between the superfamilies Psychoidea, Tineoidea and Yponomeutoidea (Lepidoptera) in the light of the functional morphology of the male genitalia. Part 1. Functional morphology of the male genitalia. Ent. Obozr., 55:533–548. [In Russian. 1977 English trans.: Ent. Rev., 55(3):19–29].
1977. Phylogenetic relationships of the superfamilies Psychoidea, Tineoidea and Yponomeutoidea (Lepidoptera) with regard to functional morphology of the male genital apparatus, Part 2. Phylogenetic relationships of the families and subfamilies. Ent. Obozr.,56:19–30. [In Russian. 1978 English trans.: Ent. Rev., 56(1):14–21].

Kyrki, J.
1984. The Yponomeutoidea: a reassessment of the superfamily and its suprageneric groups (Lepidoptera). Ent. Scand., 15:71–84.

Lameere, H.
1907. Manuel de la fauna de Belgique. III: Insectes supérieurs. Brussels: H. Lamertin. 870 pp.

Legrand, H.
1965. Lépidoptères des iles Seychelles et d' Aldabra. Mém. Mus. Natl. Hist. Nat. (Paris), Ser. A. (Zool.), 37:1–210, 16 pl.

Le Marchand, S.
1937a. Tineina. Les Glyphipterygidae. Amat. Papill., 8:189–193.
1937b. Tineina. Les Heliodinidae. Amat. Papill., 8:217–221.
1945. Les microlépidoptères.- Classification des Tineina. Rev. Fr. Lepid., 10:94–111, 125–144.

Leonard, M. D.
1928. A list of the insects of New York, with a list of the spiders and certain allied groups. Cornell Univ. Agric. Exp. Sta. Mem., 101:1–1121.

Leraut, P.

1980. Liste systématique et synonymique des lépidoptères de France, Belgique et Corse. Paris: Soc. Ent. France. 334 pp.

Lhomme, L.
1948. Catalogue des lépidoptères de France et de Belgique. Tome II. Fasc. IV. Paris. 808 pp.

Lower, O. B.
1905. New Australian Lepidoptera No. 22. Trans. Roy. Soc. So. Australia, 29:103-115.

Lycklama, H. J.
1927. Naamlijst van de Nederlandsche Microlepidoptera. 30 pp.

MacKay, M. R.
1972. Larval sketches of some microlepidoptera, chiefly North American. Mem. Ent. Soc. Can., 88:1-83.

Malloch, J. R.
1901. A list of the Tortricidae and Tineina of the Parish of Bonhill, Dumbartonshire. Ent. Mon. Mag., 37:185-188.

Mann, J.
1857. Verzeichniss der im Jahre 1853 in der Gegend von Fiume gesammelten Schmetterlinge. Ent. Monatsch. Wien, 1:139-189.

Mann, J., and A. F. Rogenhofer
1878. Zur Lepidopteren-Fauna des Dolomiten-Gebietes. Verh. Zool.-Bot. Ges. Wien, 27:491-500.

Martini, W.
1916. Verzeichnis Thüringer Kleinfalter aus den Familien Pyralidae-Micropterygidae. Dtsch. Ent. Zeit. Iris, 30: 110-144, 153-186.

Matsumura, S.
1931. 6000 illustrated insects of Japan-Empire. Tokyo. 1496 pp. (index, 191 pp.). [In Japanese].

Mayr, E.
1969. Principles of systematic zoology. New Tork: McGraw-Hill. 428 pp.

McDunnough, J. H.
1939. Checklist of the Lepidoptera of Canada and the United States of America. Part II. Microlepidoptera. Mem. So. Calif. Acad. Sci., 2:1-171.

Meyrick, E.
1880. Descriptions of Australian micro-lepidoptera. IV. Tineina. Proc. Linn. Soc. N.S.W., 5:204-271.
[1888]. Descriptions of New Zealand Tineina. Trans. New Zealand Inst., 20:77-106.
1895. A handbook of British Lepidoptera. London: Macmillan. 843 pp.

1906. On the genus **Imma**, Walk., (= **Tortricomorpha**, Feld.). Trans. Ent. Soc. Lond., 1906:169-206.

1907. Descriptions of Australasian micro-lepidoptera. Proc. Linn. Soc. N.S.W., 32:47-150.

1909. Descriptions of Indian micro-lepidoptera. IX. J. Bombay Nat. Hist. Soc., 19:410-437.

1911a. Notes and descriptions of New Zealand Lepidoptera. Trans. New Zealand Inst., 43:58-78.

1911b. Tortricina and Tineina. In, Reports of the Percy Sladen Trust Expedition to the Indian Ocean in 1905, under the leadership of Mr. J. Stanley Gardiner, M.A. V. 3. Trans. Linn. Soc. Lond., Zool., (2) 14:263-307.

1912. Glyphipterygidae. Exotic Microlep., 1:35-63.

1913a. Glyphipterygidae. Exotic Microlep., 1:67-70, 98-104.

1913b. Carposinidae, Heliodinidae, Glyphipterygidae. In, H. Wagner (ed.), Lepidopterorum Catalogus, 13:1-53. Berlin: W. Junk.

1914a. Descriptions of New Zealand Lepidoptera. Trans. New Zealand Inst., 46:101-118.

1914b. Glyphipterygidae. Exotic Microlep., 1:283-284.

1914c. Lepidoptera Heterocera. Fam. Glyphipterygidae. In, P. Wytsman (ed.), Genera Insectorum, 164:1-39, 2 pl. Tervuren, Belgium.

1915a. Descriptions of South American micro-lepidoptera. Trans. Ent. Soc. Lond., 1915:201-256.

1915b. Revision of New Zealand Tineina. Trans. New Zealand Inst., 47:205-244.

1916. Descriptions of New Zealand Lepidoptera. Trans. New Zealand Inst., 46:414-419.

1918. Glyphipterygidae. Exotic Microlep., 2:191-196.

[1920]. A sketch of our present knowledge of Indian microlepidoptera. Rep. 3rd Ent. Meeting, Pusa 1919, 3:999-1009.

1920a. Descriptions of South African micro-lepidoptera. Am. So. Afr. Mus., 17:273-318.

1920b. Glyphipterygidae. Exotic Microlep., 2:335-338.

1921a. Descriptions of South African micro-lepidoptera. Ann. Transvaal Mus., 8:49-148.

1921b. New microlepidoptera. Zool. Meded., 6:145-202.

1922. Glyphipterygidae. Exotic Microlep., 2:481-493.

1928. A revised handbook of British Lepidoptera. Marlborough, England. 914 pp. [1968 facsimile: E. W. Classey, London].

1931. Glyphipterygidae. Exotic Microlep., 4:180-185.

1935. List of microlepidoptera of Chekiang, Kiangsu and Hu-

nan. In, A. Caradja and E. Meyrick, Materialen zu einer Microlepidopteren Fauna der chinesischen Provincen Kiangsu, Chekiang und Hunan. Berlin: Friedländer. 96 pp. (Pp. 44-96).
Millière, P.
    [1876]. Catalogue raisonné des lépidoptères du département des Alpes-Maritimes [part]. Mém. Soc. Sci. Nat. Cannes, 5:251-455.
Möbius, E.
    1936. Verzeichnis der Kleinschmetterlinge von Dresden und Umgebung. Dtsch. Ent. Zeit. Iris, 50:101-134, 167-196.
Moore, F.
    1884-[87]. The Lepidoptera of Ceylon. Vol. III. London: Reeve. 578 pp., pl. 144-215. [1884:1-88, pl. 144-157. [1885]: 89-304, pl. 158-181. [1886]:305-392, pl. 182-195. [1887]: 393-578, pl. 196-215].
Moriuti, S.
    1960. Descriptions of the larva and pupa of Glyphipteryx semiflavana Issiki (Lepidoptera: Glyphipterygidae). Kontyu (Tokyo), 28:16-21. [In Japanese].
    1977. Yponomeutidae s. lat. (Insecta: Lepidoptera). In, Fauna Japonica. Tokyo: Keigaku Publ. Co. 327 pp., 96 pl.
Moriuti, S., and T. Saito
    1964. Glyphipteryx semiflavana Issiki and the allied new species from Japan (Lepidoptera: Glyphipterygidae). Ent. Rev. Japan (Tokyo), 16:60-63, pl. 9-10. [In Japanese].
Morley, C., and W. Rait-Smith
    1933. The hymenopterous parasites of the British Lepidoptera. Trans. Ent. Soc. Lond, 81:133-183.
Morris, F. O.
    1872. A natural history of British moths.   Vol. IV.   London: G. Bell & Sons. 321 pp., pl. 97-132.
Morris, F. O., and W. E. Kirby
    1896. A history of British moths. (5th ed.). Vol. IV.   London: G. Bell & Sons. 321 pp., pl. 97-132.
Morse, S. R.
    1910. The insects of New Jersey.   Ann. Rep. New Jersey St. Mus., 1909:1-888.
Möschler, B.
    1866. Aufzählung der in Andalusien 1865 von Herrn Graf v. Hoffmannsegg gesammelten Schmetterlinge.   Berliner Ent. Zeit., 10:136-146.
Munroe, E. G.
    1972-[74]. Pyraloidea. Pyralidae (part). In, R. B. Dominick, et al. (eds.), The moths of America north of Mexico.

Fasc. 13.1. London: E. W. Classey. 304 pp., 13 pl.; pl. A-K.

1976. Pyraloidea. Pyralidae (part). Ibid., Fasc. 13.2. London: E. W. Classey. 150 pp., 9 pl; pl. A-H, J-U [no pl. I].

Naumann, C. M.
1971. Untersuchungen zur Systematik und Phylogenie der holarktischen Sesiiden (Insecta, Lepidoptera). Bonner Zool. Monog. (Bonn), 1:1-190. [English trans. 1977: Studies on the systematics and phylogeny of Holarctic Sesiidae (Insecta, Lepidoptera). New Delhi: Amerind. (Smithsonian Inst., Washington). 208 pp.].

Naumann, F.
1939. Tabellarisches Verzeichnis der europäischen Lepidopteren mit reduziertem oder schwach ausgebildetem Rüssel und deren Flugzeit. Dtsch. Ent. Zeit. Iris, 52: 112-121.

Needham, J. G.
1955. Notes on a leaf-rolling caterpillar and on some of its associates. Ecology, 36:346-352.

Newman, E.
[1856]. Characters of a few Australian Lepidoptera, collected by Mr. Thomas R. Oxley. Trans. Ent. Soc. Lond., 1855: 281-300.

Nickerl, O.
1908. Beiträge zur Insekten-Fauna Böhmens (Tineen). Prague. 161 pp.

Oberberger, J.
1952. Entomologie I. Anatomie, Morfologie a Embryologia Hmyzu. Prague: Prirodovedecke. 869 pp.

Peterson, A.
1965. Larvae of insects. An introduction to Nearctic species. Part 1. Lepidoptera and plant infesting Hymenoptera. (5th ed.). Columbus, Ohio. 315 pp.

Philpott, A.
1918. Descriptions of new species of Lepidoptera. Trans. New Zealand Inst., 50:125-132.

1927. The male genitalia of the New Zealand Glyphipterygidae. Trans. New Zealand Inst., 58:337-347.

Pierce, F. N., and J. W. Metcalfe
1935. The genitalia of the tineid families of the Lepidoptera of the British Islands. An account of the morphology of the male clasping organs and the corresponding organs of the female. Warmington: England. 116 pp., 68 pl.

Porritt, G. T.

1904. List of Yorkshire Lepidoptera. Ent. Trans. Yorkshire Naturalists' Union, 2:1-269.

Powell, J. A.
1973. A systematic monograph of New World ethmiid moths (Lepidoptera: Gelechioidea). Smithson. Contr. Zool., 120:1-302.

Procter, W.
1938. Biological survey of the Mount Desert region. Part VI. The insect fauna with references to methods of capture, food plants, the flora and other biological features. Philadelphia: Wistar Inst. Anat. Biol. 496 pp.
1946. Ibid. Part VII. (rev. ed.). Philadelphia: Wistar Inst. Anat. Biol. 566 pp.

Rebel, H.
1901. II. Theil: Famil. Pyralidae-Micropterygidae. In, O. Staudinger and H. Rebel, Catalog der Lepidopteren des palaearctischen Faunengebietes. Berlin: Friedländer. 368 pp.
1916. Die Lepidopterenfauna Kretas. Ann. Kais.-Könn. Naturhist. Hofmus. (Vienna), 30:66-172, pl. 4.
1940. Die Lepidopterenfauna des azorischen Archipels. Mit 1. Anhang: eine Lepidopteren-Ausbeute von Madeira. Comm. Biol., 8:1-59, pl. 1-2.

Rennie, J.
1832. A conspectus of the butterflies and moths found in Britain, with their english and systematic names, times of appearance, sizes, colours; their caterpillars, and various localities. London: W. Orr. 287 pp.

Reutti, C. H.
1898. Übersicht der Lepidopteren-Fauna des Grossherzogtums Baden (und der anstossenden Länder). Verh. Naturwiss. Ver. Karlsruhe, 12:1-361.

Richards, O. W., and R. G. Davies
1977. Imms' general textbook of entomology. 10th ed. Vol. 2: Classification and biology. London: Chapman & Hall. Pp. 419-1354.

Riley, S. V.
1891. Tineina. In, J. B. Smith, List of the Lepidoptera of boreal America. Philadelphia: Amer. Ent. Soc. 124 pp.

Robinson, G. S.
1976. The preparation of slides of Lepidoptera genitalia with special reference to the microlepidoptera. Ent. Gaz., 27:127-132.

Rosenstock, R.
1885. Notes on Australian Lepidoptera, with descriptions of

new species. Ann. Mag. Nat. Hist., (5) 16:376-385, 421-443.

Schütze, K. T.
1902. Die Kleinschmetterlinge der sächsischen Oberlausitz. III. Theil (Tineina, Micropterygina). Dtsch. Ent. Zeit. Iris, 15:1-49.
1931. Die Biologie der Kleinschmetterlinge unter besonderer Berücksichtigung ihrer Nährpflanzen und Erscheinungszeiten. Frankfurt: Internatl. Ent. Ver. E.V. 235 pp.

Schütze, K. T., and A. Roman
1931. Schlupfwespen. Isis Budissina, 12:1-12.

Seebold, T.
[1899]. Beiträge zur Kenntnis der Microlepidopterenfauna Spaniens und Portugals. Dtsch. Ent. Zeit. Iris, 11:291-322.

Smith, John B.
[1900]. Insects of New Jersey. A list of the species occurring in New Jersey, with notes on those of economic importance. Ann. Rep. St. Board Agric., 1899:1-755.

Snellen, P. C. T.
1882. De Vlinders van Nederland. Microlepidoptera. Leiden: Brill. 1196 pp., 14 pl. (2 vol.).

Sodoffsky, W.
1837. Etymologische Untersuchungen ueber die Gattungsnamen der Schmetterlinge. Bull. Soc. Imperial Nat. Moscou, 10:76-97.

Sorhagen, L.
1885. Aus meinem entomologischen Tagebuche. Berliner Ent. Zeit., 29:81-108.

Speyer, A.
[1856]. Deutsche Schmetterlingskunde für Anfänger. Nebst einer Anleitung zum Sammeln. Mainz: Kunze. 271 pp., 34 pl.

Spuler, A.
1910. Die Schmetterlinge Europas. II. Band. Stuttgart: Schweizerbart. 523 pp.

Standfuss, M. R.
1849. Lepidopterologische Beiträge zur Kenntnis der Iserwiesen. Zeit. Ent. (Breslau), 4:19-24, Lepid. pl. 2.

Stainton, H. T.
1854a. List of the specimens of British animals in the collection of the British Museum. Part XVI. Lepidoptera. London: Br. Mus. (Nat. Hist.). 199 pp.
1854b. Insecta Brittanica. Lepidoptera: Tineina. London: Reeve. 331 pp. [Vol. 3].

1855. Habit of the larva of **Glyphipteryx haworthana**. Zoologist (london), 13:4654-4655.

1859a. New British species in 1858. Ent. Ann., 1859:145-157.

1859b. A manual of British butterflies and moths. Vol II.(comprising the slender-bodied and small moths). London: Van Voorst. 480 pp.

1867. British butterflies and moths: An introduction to the study of our native Lepidoptera. London: Reeve. 292 pp., 16 pl.

1869. Tineina of southern Europe. London: Van Voorst. 370 pp.

1870. The natural history of the Tineina. Vol. XI. London: Van Voorst. 330 pp., 8 pl.

1872. The Tineina of North America, by (the late) Dr. Brackenridge Clemens. (Being a collected edition of his writings on that group of insects). London: Van Voorst. 282 pp.

Staudinger, O.

1870. Beitrag zur Lepidopterenfauna Griechenlands. Hor.Soc. Ent. Ross., 7:2-304, 3 pl.

1879. Lepidopteren-Fauna Kleinasien's. Hor. Soc. Ent. Ross., 15:159-435.

Stephens, J. F.

1829. A systematic catalogue of British insects. Part 2: Insecta Haustellata. London: Baldwin & Cradock. 388 pp.

1834. Illustrations of British entomology; or, a synopsis of indigenous insects: containing their generic and specific distinctions; with an account of their metamorphosis, times of appearance, localities, food, and economy, as far as practicable. Haustellata. Vol. IV. London. 434 pp., pl. 33-40.

Sterneck, J., and F. Zimmermann

1933. Sterneck prodromus der Schmetterlingsfauna Böhmens. II. Teil: Microlepidoptera. Karlsbad. 168 pp.

Thompson, W. R.

1946. A catalogue of the parasites and predators of insect pests. Sect. 1, Pt. 7. Parasites of the Lepidoptera (G-M). Belleville, Canada: Imp. Parasite Serv. [= Commonwealth Inst. Biol. Control, Ottawa]. Pp. 285-359.

Tillyard, R. J.

1926. The insects of Australia and New Zealand. Sydney: Angus & Robertson. 560 pp., 41 pl.

Toll, S.

1936. Untersuchung der Genitalien bei **Pyrausta purpuralis** L. und **P. ostrinalis** Hb., nebst Beschreibung 11 neuer Microlepidopteren-Arten.   Ann. Mus. Zool. Polonici, 11: 403-413, pl. 47-49.

1956. Glyphipterygidae. In, Klucze do oznaczania owadow Polski, 27. Lepidoptera, 39:1-36.   Krakow: Polish Ent. Soc. [In Polish].

Treitschke, F.
     1833. Schaben. Geitschen. In, F. Ochsenheimer, Die Schmetterlinge von Europa. Vol. 9 [2nd pt.]. Leipzig: Fleischer. 294 pp.

Turati, E.
     1879. Contribuzione alla fauna Lepidotterologica Lombarda. Bull. Soc. Ent. Ital., 11:153-208, pl. 7-8.

Turner, A. H.
     1955. Lepidoptera of Somerset. Somerset: Somersetshire Archaeolog. Nat. Hist. Soc. 195 pp.

Turner, A. J.
     1898. Descriptions of new micro-lepidoptera from Queensland. Trans. Roy. Soc. S. Austr., 1898:200-214.

     1903. Revision of Australian microlepidoptera.   Proc. Linn. Soc. N.S.W., 28:42-92.

     1913. Studies in Australian microlepidoptera. Proc. Linn. Soc. N.S.W., 38:174-228.

     1926. Studies in Australian Lepidoptera.   Trans. Roy. Soc. S. Austr., 50:120-155.

     1927. New and little-known Tasmanian Lepidoptera. Pap. Roy. Soc. Tasmania, 1926:119-164.

     1939. A second revision of the Lepidoptera of Tasmania. Proc. Roy. Soc. Tasmania, 1938:57-115.

     1942. Fragmenta lepidopterologica. Proc. Roy. Soc. Queensland, 53:61-96.

     1947. A review of the phylogeny and classification of the Lepidoptera. Proc. Linn. Soc. N.S.W., 71:303-338.

Viette, P. E. L.
     1947. Catalogue des microlépidoptères de Madagascar et des archipels environments. Mém. Inst. Sci. Madagascar,(A) 1:31-75.

     1949. Essai d'un tableau de détermination des familles de Tinéides de la faune francaise. Rev. Fr. Lépid.,12:16-23.

Vorbrodt, C.
     1928. Die Schmetterlinge von Zermatt. Dtsch. Ent. Zeit. Iris, 42:7-130, 1 pl.

     1931. Tessiner und Misoxer Schmetterlinge. II. "Microlepidoptera." Dtsch. Ent. Zeit. Iris, 45:91-140.

Walker, F.
     1864. List of the specimens of lepidopterous insects in the

collection of the British Museum. Part XXX. Tineites. London: Br. Mus. Pp. 837-1096.

Walsingham, T. de Grey

1881. On some North-American Tineidae.　Proc. Zool. Soc. Lond., 1881:301-325, pl. 35-36.

1897a. Western equatorial African micro-lepidoptera. Trans. Ent. Soc. Lond., 1897:33-67, pl. 2-3.

1897b. Revision of the West-Indian micro-lepidoptera, with descriptions of new species. Proc. Zool. Soc. Lond., 1897:54-183.

[1908]. Microlepidoptera of Tenerife. Proc. Zool. Soc. Lond., 1907:911-1034, pl. 51-53.

1914. Hemerophilidae. In, Tineina, Pterophorina, Orneodina and Pyralidina and Hepialina (part). In, F. D. Godman and O. Salvin (eds.), Biologia Centrali-Americana.[Vol. 42]. Insecta. Lepidoptera-Heterocera. Vol. 4. London. 482 pp., 9 pl. (Pp. 300-319).

Waters, E. G. R.

1928. Observations on Glyphipteryx schoenicolella Boyd. Ent. Mon. Mag., 64:252-253.

Watson, A., and P. E. S. Whalley

1975. The dictionary of butterflies and moths in color.　New York: McGraw-Hill. 296 pp.

Watt, M. N.

1920. The leaf-mining insects of New Zealand.　Trans. New Zealand Inst., 52:439-446, pl. 30.

Westwood, J. O.

[1838]-40a. Synopsis of the genera of British insects. London. 158 pp. [[1838]:1-48. [1839]:49-80. 1840:81-158.]

[1839]-40b. An introduction to the modern classification of insects; founded on natural habits and corresponding organization of the different families. Vol. II. London: Longman. 587 pp. [[1839]:pt. IX-XIV.　1840:pt. XV-XVI.]

1845. In, H. N. Humphreys and J. O. Westwood, British moths and their transformations. Vol. 2. London: W. Smith. 268 pp., pl. 57-123.

1854. Index entomologicus; or, a complete illustrated catalogue, consisting of upwards of two thousand accurately coloured figures of the lepidopterous insects of Great Britain. Rev. ed. [of Wood, 1833-1839]. London: Willis. 259 pp., 59 pl.

Wocke, M. F.

1871. Microlepidoptera. In, O. Staudinger and M. F. Wocke, Catalog der Lepidopteren des europaeischen Faunengebietes. Dresden. 426 pp.

[1876]. Zweite Abtheilung. Kleinschmetterlinge. Band II. Die Motten und Federmotten. Heft II. In, H. von Heinemann and M. F. Wocke, Die Schmetterlinge Deutschlands und der Schweiz. Braunscheig: Schwetschke. Pp. 389–825.

Wood, W.

1833-39. Index entomologicus; or, a complete illustrated catalogue, consisting of 1944 figures, of the lepidopterous insects of Great Britain. London. 266 pp., 54 pl. [1837: 167-214, pl. 39-47.]

Wörz, A.

1954. Die Lepidopterenfauna von Württemberg. II. Microlepidoptera. Kleinschmetterlinge. Jahresheft. Ver. Naturk. Württemberg, 109:83-130.

Wu, C. F.

1938. Catalogus insectorum sinensium. (Catalogue of Chinese insects). Vol. 4. [Lepidoptera]. Peking: Fan Mem. Inst. Biol. 1007 pp.

Zeller, P. C.

1839a. Versuch einer naturgemässen Eintheilung der Schaben. Isis von Oken, 32:167-220.

1839b. Kritische Bestimmung der in Degeers Abhandlungen zur Geschichte der Insekten enthaltenen Lepidopteren. Isis von Oken, 32:243-348.

1847. Bemerkungen über die auf einer Reise nach Italien und Sicilien beobachteten Schmetterlingsarten. VII. Isis von Oken, 40:641-673.

1877. Exotische Microlepidoptera. Hor. Soc. Ent. Ross., 13: 3-493, 9 pl.

Zerny, H.

1927. Die Lepidopterenfauna von Albarracin in Aragonien. EOS Rev. España Ent., 3:299-487, pl. 8-10.

Zetterstedt, J. W.

[1839]-40. Insecta Lapponica. Lipsiae: Voss. 1140 pp.[[1839]: 1-1014. 1840:1015-1140.]

INDEX TO PLANT NAMES

INDEX TO ANIMAL NAMES

T - #0516 - 101024 - C0 - 229/152/14 - PB - 9780916846329 - Gloss Lamination